READINGS in the Marriage and Family Experience

Intimate Relationships in a Changing Society

BRYAN STRONG
University of California, Santa Cruz

CHRISTINE DeVAULT
Cabrillo College

BARBARA W. SAYAD
California State University, Monterey Bay

 Wadsworth Publishing Company
I(T)P® An International Thomson Publishing Company

Belmont, CA • Albany, NY • Bonn • Boston • Cincinnati • Detroit • Johannesburg • London
Madrid • Melbourne • Mexico City • New York • Paris • Singapore • Tokyo • Toronto • Washington

Sociology Editor: Denise Simon
Marketing Manager: Chaun Hightower
Editorial Assistant: Angela Nava
Cover Design: Margarite Reynolds
Cover: Original quilt design, "Address Unknown," by artist Freddy Moran
Compositor: Lorrain and Jeff Sargent/Pacific Publications
Printer: Globus Printing & Packaging

COPYRIGHT © 1998 by Wadsworth Publishing Company
A Division of International Thomson Publishing Inc.
I(T)P® The ITP logo is a registered trademark under license.

Printed in the United States of America
7 8 9 10

For more information, contact Wadsworth Publishing Company, 10 Davis Drive, Belmont, CA 94002, or electronically at http://www.thomson.com/wadsworth.html

International Thomson Publishing Europe
Berkshire House 168-173
High Holborn
London, WC1V 7AA, England

Thomas Nelson Australia
102 Dodds Street
South Melbourne 3205
Victoria, Australia

Nelson Canada
1120 Birchmount Road
Scarborough, Ontario
Canada M1K 5G4

International Thomson Publishing GmbH
Königswinterer Strasse 418
53227 Bonn, Germany

International Thomson Editores
Campos Eliseos 385, Piso 7
Col. Polanco
11560 México D.F. México

International Thomson Publishing Asia
221 Henderson Road
#05-10 Henderson Building
Singapore 0315

International Thomson Publishing Japan
Hirakawacho Kyowa Building, 3F
2-2-1 Hirakawacho
Chiyoda-ku, Tokyo 102, Japan

International Thomson Publishing
Southern Africa
Building 18, Constantia Park
240 Old Pretoria Road
Halfway House, 1685 South Africa

All rights reserved. No part of this work covered by the copyright hereon may be reproduced or used in any form or by any means—graphic, electronic, or mechanical, including photocopying, recording, taping, or information storage and retrieval systems—without the written permission of the publisher.

ISBN-13: 978-0-534-53763-0
ISBN-10: 0-534-53763-4

Contents

PREFACE vii

UNIT I MEANINGS OF MARRIAGE AND FAMILY

Chapter 1 The Meaning of Marriage and Family

    1   The Fifties Family: The Way We Never Were by Stephanie Coontz 1

    2   The American Family: A Poem by Marilyn Inhinger-Tallman 5

Chapter 2 Studying Marriage and the Family

    3   Women Abandoning Work for Home: A Media Myth by Margaret L. Usdansky 7

    4   Values in the Study of Culturally Diverse Families by Peggye Dilworth-Anderson, Linda M. Burton, and William L. Turner. 9

Chapter 3 Contemporary Gender Roles

    5   Ethnicity and Gender Stereotypes by Hilary M. Lips 12

    6   The Male Mystique by Andrew Kimbrell 16

UNIT II INTIMATE RELATIONSHIPS

Chapter 4 Friendship, Love, and Commitment

    7   Making Love Work by Robert Sternberg 19

    8   On Jealousy by Leo Buscaglia 22

Chapter 5 Communication and Conflict Resolution

    9   How to Fight Fair by Judith Viorst 24

   10   Family Secrets by David Gelman and Debra Rosenberg 28

Chapter 6 Pairing and Singlehood

   11   Wired Love: Courtship in the Age of E-Mail by Lisa Napoli 32

   12   The Dilemma of Black Single Women by Gloria Naylor 34

Chapter 7   Understanding Sexuality

13	The Slumber Party by Marilyn Miller	36
14	What Is Normal? by Marty Klein	39
15	Infidelity and the Science of Cheating by Sharon Begley	41

UNIT III   FAMILY LIFE
Chapter 8   Pregnancy and Childbirth

16	A Native American Birth Story by Marcie Rendon	44
17	There is No Good Decision by Nancy R. Nerenberg	47

Chapter 9   Marriage as Process: Family Life Cycles

18	Gay Families Come Out by Barbara Kantrowitz	50
19	Breaking Parental Ties by Jean Marzollo	54
20	At Grandma's Table by Anne McCarroll	57

Chapter 10   Parents and Children

21	On Not Having Kids by Jon Hubner	59
22	Sharing the Baby by Joanne Kates	61
23	The Lament of the Older Parent by Laurie DeVault	63

Chapter 11   Marriage, Work, and Economics

24	There's No Place Like Work by Arlie Russell Hochschild	65
25	Beating the Clock by Barbara Kantrowitz	70
26	The Feminization of Poverty by Marjorie E. Starrels, Sally Bould, and Leon J. Nicholas	72

UNIT IV   FAMILY CHALLENGES AND STRENGTHS
Chapter 12   Families and Wellness

27	What Pushes Your Stress Button? by Anne Cassidy	74
28	I Want to Die at Home by Anne Ricks Sumers	78
29	Rules in Alcoholic Families by Sharon Wegscheider	80

Chapter 13 Family Violence and Sexual Abuse

   30   The Deadly Rage of a Battered Wife by Janet Bukovinsky 83

   31   Breaking the Silence by Meredith Maran 86

   32   Marriage or Rape? by Peter Annin and Kendall Hamilton 89

Chapter 14 Coming Apart: Separation and Divorce

   33   Children of Divorce: Ten Years After by Judith Wallerstein 91

   34   When Dads Disappear by David Popenoe 94

Chapter 15 New Beginnings: Single-Parent Families and Stepfamilies

   35   Friends Through It All by Elizabeth Stark 97

   36   Single Mothers and Helping Kin in an African-American Community by Carol Stack 100

Chapter 16 Marriage and Family Strengths

   37   Of Course You Love Them—But Do You Enjoy Them? by Anita Shreve 104

   38   Kwanzaa: A "New" Tradition for African Americans by Eric V. Copage 108

Preface

This reader is intended to expand the study of intimate relationships by providing a selection of both personal and intellectual perspectives. Chosen from a variety of sources and disciplines, the articles will challenge and broaden the student's understanding of marriages, families, and other close relationships and will provide the instructor a ready source of stimulating and discussion-provoking supplemental materials. We introduce this first anthology of readings with the hope that it engages students' minds and enriches their study of marriage and family.

The selection of 38 readings comes from essays, articles, and excerpts from books, magazines, and newspapers. With a desire to continue to assist students to see beyond their own experiences, we reviewed and chose articles in light of their ability to provide a cross-cultural, multi-disciplinary, and/or personal approach to a topic. To further assist this process, we created "reflection" (critical thinking) questions for each reading. These can be used as discussion starters, personal introspections, or essay questions.

We hope that you find this collection to be an effective supplement to the study of marriage and family as well as a tool to help bring to light the personal aspects which give this field its meaning and life.

Acknowledgments

The editors and Wadsworth Publishing Company wish to express their sincere gratitude to the many authors, journals, publishers that permitted us to reprint their articles. We are indebted to senior editor Denise Simon for supporting this project and to editorial assistant Angela Nava.

About the Editors

Bryan Strong published the first edition of *The Marriage and Family Experience* in 1979. He taught human sexuality at the University of California Santa Cruz for many years, after receiving his doctorate from Stanford University in 1972.

Christine DeVault has co-authored *The Marriage and Family Experience* since its second edition in 1983. She is a Certified Family Life Educator and teaches courses in relationships at Cabrillo College in Aptos, California.

Barbara W. Sayad teaches courses in human sexuality, women's issues, and health at California State University, Monterey Bay, and is co-author of the 7th edition of *The Marriage and Family Experience* with Bryan Strong and Christine DeVault.

According to historian Stephanie Coontz, images of the traditional family exert a powerful emotional pull on many Americans. Coontz argues that the fifties family and TV reruns are the source of our images of the traditional family. She also argues that our image of the 1950s family is incomplete because it overlooks many of the negative aspects of fifties family life.

The Fifties Family: The Way We Never Were

Excerpted from *The Way We Never Were*
By Stephanie Coontz

The "traditional" family of the 1950s was a qualitatively new phenomenon. At the end of the 1940s, all the trends characterizing the rest of the twentieth century suddenly reversed themselves: For the first time in more than one hundred years, the age for marriage and motherhood fell, fertility increased, divorce rates declined, and women's degree of educational parity with men dropped sharply. In a period of less than ten years, the proportion of never-married persons declined by as much as it had during the entire previous half century.

People married at a younger age, bore their children earlier and closer together, completed their families by the time they were in their late twenties, and experienced a longer period living together as a couple after their children left home. The traditional range of acceptable family behaviors—even the range in the acceptable number and timing of children—narrowed substantially.

The values of 1950s families also were new. The emphasis on producing a whole world of satisfaction, amusement, and inventiveness within the nuclear family had no precedents. Historian Elaine Tyler May comments: "The legendary family of the 1950s...was not, as common wisdom tells us, the last gasp of 'traditional' family life with deep roots in the past. Rather, it was the first wholehearted effort to create a home that would fulfill virtually all its members' personal needs through an energized and expressive personal life."

Beneath a superficial revival of Victorian domesticity and gender distinctions, a novel rearrangement of family ideals and male-female relations was accomplished. For women, this involved a reduction in the moral aspect of domesticity and an expansion of its orientation toward personal service. Nineteenth-century middle-class women had cheerfully left housework to servants, yet 1950s women of all classes created makework in their homes and felt guilty when they did not do everything themselves. The amount of time women spent doing housework actually increased during the 1950s, despite the advent of convenience foods and new, labor-saving appliances; child care absorbed more than twice as much time as it had in the 1920s. By the mid-1950s, advertisers' surveys reported on a growing tendency among women to find "housework a medium of expression for...[their] femininity and individuality."

Acceptance of domesticity was the mark of middle-class status and upward mobility. In sitcom families, a middle-class man's work was totally irrelevant to his identity; by the same token, the problems of working-class families did not lie in their economic situation but in their failure to create harmonious gender roles. Working-class and ethnic men on television had one defining characteristic: They were unable to control their wives. The families of middle-class men, by contrast, were generally well behaved.

Not only was the 1950s family a new invention; it was also a historical fluke, based on a unique and temporary conjuncture of economic, social, and political factors. During the war, Americans had saved at a rate more than three times higher than that in the decades before or since. Their buying power was further enhanced by America's extraordinary competitive advantage at the end of the war, when every other industrial power was devastated by the experience. This privileged economic position sustained both a tremendous expansion of middle-class management occupations and a new honeymoon between management and organized labor: During the 1950s, real wages increased by more than they had in the entire previous half century.

Poverty, Diversity, and Social Change

Even aside from the exceptional and ephemeral nature of the conditions that supported them, 1950s family strategies and values offer no solution to the discontents that underlie contemporary romanticization of the "good old days." The reality of these families was far more painful and complex than the situation-comedy reruns or the expurgated memories of the nostalgic would suggest. Contrary to popular opinion, "Leave It to Beaver" was not a documentary.

In the first place, not all American families shared in the consumer expansion that provided Hotpoint appliances for June Cleaver's kitchen and a vacuum cleaner for Donna Stone. A full 25 percent of Americans, forty to fifty million people, were poor in the mid-1950s, and in the absence of food stamps and housing programs, this poverty was searing. Even at the end of the 1950s, a third of American children were poor. Sixty percent of Americans over sixty-five had incomes below $1,000 in 1958, considerably below the $3,000 to $10,000 level considered to represent middle-class status. A majority of elders also lacked medical insurance. Only half the population had savings in 1959; one-quarter of the population had no liquid assets at all. Even when we consider only native-born, white families, one-third could not get by on the income of the household head.

In the second place, real life was not so white as it was on television. Television, comments historian Ella Taylor, increasingly ignored cultural diversity, adopting "the motto 'least objectionable programming,' which gave rise to those least objectionable families, the Cleavers, the Nelsons and the Andersons." Such families were so completely white and Anglo-Saxon that even the Hispanic gardener in "Father Knows Best" went by the name of Frank Smith. But contrary to the all-white lineup on the television networks and the streets of suburbia, the 1950s saw a major transformation in the ethnic composition of America. More Mexican immigrants entered the United States in the two decades after the Second World War than in the entire previous one hundred years. Prior to the war, most blacks and Mexican-Americans lived in rural areas, and three-fourths of blacks lived in the South. By 1960, a majority of blacks resided in the North, and 80 percent of both blacks and Mexican-Americans lived in cities. Postwar Puerto Rican immigration was so massive that by 1960 more Puerto Ricans lived in New York than in San Juan.

These minorities were almost entirely excluded from the gains and privileges accorded white middle-class families. The June Cleaver or Donna Stone homemaker role was not available to the more than 40 percent of black women with small children who worked outside the home. Twenty-five percent of these women headed their own households, but even minorities who conformed to the dominant family form faced conditions quite unlike those portrayed on television. The poverty rate of two-parent black families was more than 50 percent, approximately the same as that of one-parent black ones. Migrant workers suffered "near-medieval" deprivations, while termination and relocation policies were employed against Native Americans to get them to give up treaty rights.

African Americans in the South faced systematic, legally sanctioned segregation and pervasive brutality, and those in the North were excluded by restrictive covenants and redlining from many benefits of the economic expansion that their labor helped sustain. Whites resisted, with harassment and violence, the attempts of blacks to participate in the American family dream.

More Complexities: Repression, Anxiety, Unhappiness, and Conflict

In the 1947 bestseller, *The Modern Woman: The Lost Sex,* Marynia Farnham and Ferdinand Lundberg described feminism as a "deep illness," called the notion of an independent woman a "contradiction in terms," and accused women who sought educational or employment equality of engaging in symbolic "castration" of men. As sociologist David Riesman noted, a woman's failure to bear children went from being "a social disadvantage and sometimes a personal tragedy" in the nineteenth century to being a "quasi-perversion" in the 1950s. The conflicting messages aimed at women seemed almost calculated to demoralize: At the same time as they labeled women "unnatural" if they did not seek fulfillment in motherhood, psychologists and popular writers insisted that most modern social ills could be traced to domineering mothers who invested too much energy and emotion in their children. Women were told that "no other experience in life...will provide the same sense of fulfillment, of happiness, of complete pervading contentment" as motherhood. But soon after delivery they were asked, "Which are you first of all, Wife or Mother?" and warned against the tendency to be "too much mother, too little wife."

All women, even seemingly docile ones, were deeply mistrusted. They were frequently denied the right to serve on juries, convey property, make contracts, take out credit cards in their own name, or establish residence. A 1954 article in *Esquire* called working wives, a "menace"; a *Life* author termed married women's employment a "disease." Women were excluded from several professions, and some states even gave husbands total control over family finances. There were not many permissible alternatives to baking brownies, experimenting with new canned soups, and getting rid of stains around the collar.

Men were also pressured into acceptable family roles, since lack of a suitable wife could mean the loss of a job or promotion for a middle-class man. Bachelors were categorized as "immature," "infantile," "narcissistic," "deviant," or even "pathological." Family advice expert Paul Landis argued: "Except for the sick, the badly crippled, the deformed, the emotionally warped and the mentally defective, almost everyone has an opportunity [and, by clear implication, a duty] to marry."

Between one-quarter and one-third of the marriages contracted in the 1950s eventually ended in divorce; during that decade two million legally married people lived apart from each other. Many more couples simply toughed it out. Sociologist Mirra Komarovsky concluded that of the working-class couples she interviewed in the 1950s, "slightly less than one-third [were] happily or very happily married."

National polls found that 20 percent of all couples considered their marriages unhappy, and another 20 percent reported only "medium happiness."

A successful 1950s family, moreover, was often achieved at enormous cost to the wife, who was expected to subordinate her own needs and aspirations to those of both her husband and her children. In consequence, no sooner was the ideal of the postwar family accepted than observers began to comment perplexedly on how discontented women seemed in the very roles they supposedly desired most. In 1949, *Life* magazine reported that "suddenly and for no plain reason" American women were "seized with an eerie restlessness." Under a "mask of placidity" and an outwardly feminine appearance, one physician wrote in 1953 there was often "an inwardly tense and emotionally unstable individual seething with hidden aggressiveness and resentment."

Although Betty Friedan's bestseller *The Feminine Mystique* did not appear until 1963, it was a product of the 1950s, originating in the discontented responses Friedan received in 1957 when she surveyed fellow college classmates from the class of 1942. The heartfelt identification of other 1950s women with "the problem that has no name" is preserved in the letters Friedan received after her book was published, letters now at the Schlesinger Library at Radcilffe.

Men tended to be more satisfied with marriage than were women, especially over time, but they, too, had their discontents. Even the most successful strivers after the American dream sometimes muttered about "mindless conformity." The titles of books such as *The Organization Man*, by William Whyte (1956), and *The Lonely Crowd*, by David Riesman (1958), summarized a widespread critique of 1950s culture. Male resentments against women were expressed in the only partly humorous diatribes of *Playboy* magazine (founded in 1953) against "money-hungry" gold diggers or lazy "parasites" trying to trap men into commitment.

People who romanticize the 1950s, or any model of the traditional family, are usually put in an uncomfortable position when they attempt to gain popular support. The legitimacy of women's rights is so widely accepted today that only a tiny minority of Americans seriously propose that women should go back to being full-time housewives or should be denied educational and job opportunities because of their family responsibilities. Yet when commentators lament the collapse of traditional family commitments and values, they almost invariably mean the uniquely female duties associated with the doctrine of separate spheres for men and women.

Reflections

1. What are some of the negative and positive aspects of fifties family life?

2. How does the family with whom you were raised compare with those of the fifties?

3. When it is time to establish your own family, which values and traditions might you choose to incorporate into your family?

2

In this poem, the author raises a number of provocative questions in helping us to define and re-define family. She effectively counters the family structure with the interactions of the family to help us see it in a new light.

The American Family: A Poem

By Marilyn Inhinger-Tallman

THE ASSIGNMENT: Write a short piece on the state of the American family.

The state...
    A set of circumstances
    of attributes characterizing...a
    thing
    at a given time...

of the American...
    America...
    God shed Grace
     crown thy good...brotherhood
    from sea to sea
    she gets your teeming masses

family...
    Fellowship
     a group of individuals living under
    one roof...
    the basic unit in society...

The state of the family—

Which family?
    The teenage mother?
    Who finds her life
    irreversibly changed
    with the cry of her newborn

    They are increasing in number...

Which family?
    The yuppie couple?
    Who give away
     their child's toys to make room for
    more.
    The dual-career couple?
    who learn Lamaze
    and breathe together
    as they bring their to-be-privileged
    preciously conceived
    one-of-a-kind
    child into the world

    They are increasing in number...

Which family?
    The cohabiting couple?
    who live with her young child
    and include his on weekends
    if his ex doesn't give him grief
    and spoil their plans

    They are increasing in number...

Which family?
    The elderly couple?
    Who live a mile
     from their married daughter and
    her kids
    and son-in-law
     they help out as much as they can
    babysitting, things like that
    for they love them so
    But what scares them most
    is that son
    got divorced last year
    and now ex-daughter-in-law
    remarries next month
    and is taking grandson Zack away
    Zack, who looks just like gran'pa
    Oh Zack
    are you lost forever?
    They are increasing in number...

Which family?
   those in the working class?
   the super rich?
   the underclass?
   what happened to the middle class?

   They are decreasing in number...

Which family?
   The dual-career couple?
   who live apart for now
   because she got a terrific promotion
   but had to move to Cleveland
   will they ever have those children
   they planned?

Which family?
   The one that lives
   in the cute house behind the white
   picket fence
   with the apricot and apple tree in
   the side yard
   in suburbia
   the one with the stay-at-home mom
   and hard-working dad
   and 2.3 kids and a dog named
   Arnie

   They are decreasing in number...

Which family?
   The abusive family?
   what about those families
   in which fathers
   drunk or sober
   creep up stairs in the dark
   to fondle the small breasts and
   feel the private parts
   of daughter
   frightened, scared to breathe
while mothers
   absent, or too dependent, or too
passive
turn their heads

and mothers, fathers, frustrated
scream
push
hit
the children
then children hit children
and later, husbands and wives and
children

Can they be increasing in
number...

The state of the family
changing
in the land of abundance
and resources
teeming masses living in families

families in different sets of circumstances
possessing different attributes
all in America
God shed grace on thee

Reflections

1. In your opinion, which is more important: the family's structure or the interactions of its members?

2. Would you describe the poem as optimistic, pessimistic, or realistic in its description of families?

3. Write a short piece on the state of the American family.

3

Much of what we know about marriages and families is derived from the media. In this reading, the statistics underlying recent headlines about working women trading jobs for marriage are examined. When reviewing the statistics cited in this article, be aware of how some might be misleading.

Women Abandoning Work for Home: A Media Myth

By Margaret L. Usdansky

"Working Women: Goin' Home," proclaimed *Barron's*.

"Superwoman has had enough," the Detroit News declared. "Young women may trade jobs for marriage," *The Wall Street Journal* said.

To hear the media tell it, large numbers of U.S. women have tired of juggling career and family. Frustrated with trying to have it all, they are returning home to husbands and kids.

There's only one hitch: It doesn't appear to be true.

"It is difficult to find any evidence that women might be leaving the labor force in large numbers to take up homemaker roles," said a recent Bureau of Labor Statistics report.

In fact, the small decline in work that got so much attention is mainly among women under 25—most of whom are neither married nor mothers.

And rather than returning to traditional gender roles, young women seem to be postponing work to extend their educations—a trend that has been equally strong, though less noticed, among young men.

The report by economist Howard Hayghe was inspired by a March cover story in *Barron's*, which predicted a "return to the '50s" with a photo of an apron-clad woman toting a pie.

Says Hayghe, "If American women were giving up their careers, I'd expect that there would be a downward trend in women's labor-force participation and there isn't."

What has happened is a slowdown in the rate at which women's work-force participation is rising.

Between 1962 and 1990, the percentage of women who worked part or full time rose each year, climbing from 37.9 percent to 57.5 percent.

But in 1991, while the nation was in a recession, the percentage of women who work dipped—less than a point—before rising again to 57.9 percent in 1993.

The decline is tied to a drop in work among very young women. The percentage of women age 20 to 24 in the work force fell from 73 percent in 1987 to 70.4 percent in 1991 before rebounding to 71.3 percent in 1993. Few economists believe that means young women are embracing domesticity. Reasons:

- The percentage of men 20 to 24 in the labor force declined, too, slipping from 85.2 percent in 1987 to 83.1 percent in 1993 with no rebound so far.
- Most of the declines among women and men under 25 seems to be explained by even larger increases in the percentage of young people in school.
- For most women under 25, "there's no home to go home to," Hayghe says. Just 30 percent of women 20 to 24 and under 10 percent of women 16 to 19 are married or have children.

The more likely explanations for lower work rates are that women are going to school longer and that the recession hurt the job market, Hayghe says.

And having more education is likely to make women more dedicated to careers, not less, because it raises their pay and leads to better jobs.

Margaret L. Usdansky, "Women Going Back Home: A Bogus Trend?" *USA Today*, © 1994. Reprinted with permission.

Says University of Maryland sociologist Harriet Presser: "If you have a good job, you get very committed."

University of Chicago junior Nicole Jordan, 20, knows she will work—even if she isn't certain in which field. She and most of her friends plan to attend graduate school and then work before they marry and have children.

"I don't see myself as the homemaker type," the math and economics major says. "I want to have that career, money aside."

For some women, working is a necessity. Often a middle-class lifestyle demands two paychecks. And with high rates of divorce and young people waiting longer to marry, many women must support themselves.

Celeste Burk, 21, decided to become the first member of her family to finish college after watching her mother struggle to make ends meet after her parents' divorce. This fall, Burk, who just earned an associate degree from Amarillo College, will head to Texas Tech University to study marketing.

"I saw how bad my mother had to struggle," Burk says, "I just want to be able to take care of myself and not depend on anybody else."

Work among single women like Burk and Jordan has risen gradually since the turn of the century. But the most rapid increase began after World War II, as large numbers of married women joined the work force.

First, mothers with children in high school went back to work, says Northwestern University economist Rebecca Blank. Then, women with children in grade school launched careers. Finally, during the 1980s, mothers of toddlers and infants joined the labor force. Today, 54 percent of women with children under 1 work outside the home.

The real question, most academics say, isn't whether large numbers of women are likely to leave the work force but whether women's work force participation will climb higher—closer to the 75 percent of men who work.

Many academics believe the percentage of working women will grow because of gains in education and wages.

Even if women's work force participation doesn't rise much, women's work experiences are likely to change in other ways. During the 1980s, women saw the first substantial narrowing of the pay gap with men. The percentage of full-time working women continues to grow steadily, rising from 45 percent in 1980 to 54 percent in 1992.

Some experts say reports say more about the wishful thinking of reporters than reality

"If people had noticed the decline in young men's labor force participation, would they have run off and said it was home and children?" asks Washington, D.C., economist Martha Farnsworth Riche, answering her own question with a loud "No."

Reflections

1. What kinds of caution should you use when reading statistics reported by the media?

2. How has the media's portrayal of woman affected you and the family in which you were raised?

3. As you read or listen to the media, what precautions should you take before accepting their conclusions?

In this excerpt from the journal Family Relations, *the authors argue that the values held by scholars affect their interpretation of ethnic families. As you read the article, pay attention to how your experiences with members of other ethnic groups have affected your perceptions of family.*

Values in the Study of Culturally Diverse Families

By Peggye Dilworth-Anderson, Linda M. Burton, and William L. Turner

The issue of values for the researcher studying culturally diverse families includes understanding the link between theorists' and researchers' value orientations regarding the family and how they chose to conceptualize and study the family. Boss, Doherty, LaRossa, Schumm, and Steinmetz (1993) suggested, "All the traditional content areas of family science, such as marital stability and parent-child socialization, as well as more recent areas such as gender relations and family violence, are thoroughly imbued with scholars' personal, cultural, and religious values" (p. 25). Further, as Coleman (1993) notes, values are seen in the scientific questions researchers ask as well as how they interpret the information gathered on families. The values of researchers also reflect American mainstream value orientations at particular sociohistorical times and the value vocabulary underlying their views. Additionally, as uncomfortable as it may be to some researchers, the study of ethnic families must involve researchers' recognition of their objective reality based on their socioeconomic and political status as well as the subjective reality of value frameworks or paradigms influencing their scientific investigations.

Several contrasting studies and findings in the literature on black families serve as illustrations of how different value orientations and world views influenced research on black families. Such contrasting views of black families are evident in studies conducted over 50 years ago, as seen in the works of Frazier (1932), who suggested that the black culture had little, if any, African characteristics. Herskovits (1935), on the other hand, disagreeing with the approach taken by Frazier, identified a number of African cultural characteristics among U.S. blacks. In the last 20 years, especially during the 1970s, when black scholars were challenging the paradigms and value orientations that influenced studies on black families, contrasting views about black families emerged (Boss et al., 1993).

These contrasting views are evident in several studies in the literature. The works of Daniel Moynihan (1965), a white scientist, and Robert Hill (1972), a black scientist using the same U.S. Census data and similar methodology, arrived at contrasting conclusions about and recommendations for black families. Moynihan described the deterioration and dysfunctionality of black families and recommended social policies that would encourage changes to reflect more mainstream ways of functioning. Hill observed resilience and strength in black families and recommended social policies that would build on their strengths.

These contrasting interpretations of black families suggest that each scientist used different value orientations and frames of reference to interpret data on black families. Moynihan assumed the underlying values inherent in the pathological or order model, while Hill's views reflected assumptions from the cultural variant or cultural relativity models (Allen, 1978; Dodson, 1988; Johnson, 1978, 1988; Staples, 1971).

The influence of social change to help create a new orientation to study families is reflected in the changing societal norms that affected the study of black families in relation to the civil rights movement of the 1960s. This movement, which described black as beautiful and acceptable rather than unacceptable, changed perceptions about and insights into blacks and other minorities in American society. Before the 1960s, black families were primarily viewed as pathological because they deviated from the norms established by models of mainstream white families (Wilson, 1986). After the 1960s, a decade when blacks in the United States demanded more rights and freedoms from society and established a more culturally distinct racial and ethnic identity, positive perceptions about them were created (Johnson, 1988). These perceptions, held by both blacks and whites, influenced how blacks and other minority families were studied. Thus, the pathological model that was formerly used as the major framework for understanding black family life was replaced with models that focused on adaptation and cultural variation, without necessarily being deviant.

Creating a new orientation that involves the inclusion of ethnic groups in empirical investigations of families challenges researchers to rethink the value they place on who should be studied and how they should be studied. The majority of the research findings in the family science literature excludes or fails to include diverse cultural groups. This lack of representation limits the development and expansion of knowledge on a large portion of families and fails to reflect the cultural and ethnic diversity in this society. Because ethnic minorities, in particular, are seldom included in family investigations, the ability to utilize comparative methodologies to understand differences both within and between groups is limited (Butler, 1984). Therefore, no insight is gained regarding the range of family structures and characteristics. These limitations reflect what DuBois (1899) called the "narrowing of the field." In this instance, this narrowing limits the development of conceptual frameworks and theories that can be used on families in general.

However, when more diverse cultural groups are included in studies of the family, researchers will be challenged with the task of reframing, expanding, and developing different ideological conceptions of the family to reflect the diversity of cultural groups in this society. Any change that takes place will first reflect a change in value orientation. Such change will foster different perceptions and attitudes toward what may be considered important in the study of the family. Further, a change in value orientation will allow for creating an ethnic reality in one of the most basic concepts in family research, the definition of the family.

Creating a new orientation in the study of the family also involves defining the family within a culturally relevant framework. The nuclear family structure has served as the dominant model for researchers studying the family. Although applicable to some families, the nuclear family structure fails to represent the majority of family structures found among ethnic groups that have an increasing number of single-parent and miltigenerational structures.

The idea of what constitutes a family and its positive characteristics needs to be expanded. This expansion will allow for the culturally relevant descriptions, explanations, and interpretations that exist among both minority and nonminority families. More specifically, such thinking would allow for understanding the ethnic reality of blacks and Native Americans as well as other ethnic/racial minority groups who have distinct extended families. It is this distinct nonnuclear existence of the family for many groups that serves as a basis for their cultural definition of the individual and family.

For example, in ethnographic studies of black families, some researchers (Aschenbrenner, 1975; Gutman, 1976; Herskovits, 1935; Jarrett, 1990; Martin & Martin, 1978; Stack, 1974) have found that relations between *fictive kin* (nonblood kin who relationally define themselves as family) are as strong and lasting as those established by blood. Adults and children who are not related by blood frequently interact with one another and are a part of the mutual aid system in the extended family. Among Native Americans, the family includes everyone in the tribal group or community. Regardless of whether there is a blood relationship, the relationships are close. Red Horse (1980) states that even when nuclear family units are found in the Native-American community, they are a part of a kinship system which has an extended structure that includes several households of kin and nonkin.

If new ways of thinking about families would address social change issues, the inclusion of ethnic/racial minority groups in studies, and relevant definitions of concepts, then it would follow that a new set of conceptual assumptions about the family would arise. These newly derived assumptions are likely to be more culturally sensitive and diverse. This cultural sensitivity is likely to challenge established ways of thinking, and most importantly, the value orientations of researchers that influence what they study about families and how they study families.

Reflections

1. Do you believe that the values held by scholars affect their interpretation of ethnic families? If so, can you cite specific examples?

2. What are some of the ways in which your own values affect your understanding of culturally diverse families?

3. What can you do to alter some of your negative views of other families?

5

Stereotyping of individuals can proceed on the basis of race, age, religion, height, social class, or any other distinction that can be used to divide people into groups and separate them from one another. In this article, the author contends that stereotyping and prejudice involve dynamic processes, not just static collections of beliefs and evaluations.

Ethnicity and Gender Stereotypes

Excerpted from *Sex and Gender, An Introduction*.
By Hilary M. Lips

When someone completing a questionnaire to measure male female stereotyping is asked about the "typical" woman or man, what kind of person do you suppose comes to the individual's mind? An elderly black woman? A middle aged Native American man confined to a wheelchair? A young woman who has trouble finding stylish clothing because she weighs two hundred pounds? Most likely, the image is influenced by the person's tendency to define "typical" with reference to himself or herself and to the people most visible in the environment. Probably, as Hope Landrine (1985) suggests, research participants cannot imagine a woman (or man) "without attributing a race, a social class, an age, and even a degree of physical attractiveness to the stimulus" (p. 66). The image, at least among the North American college students whose responses have formed much of the basis for research on gender stereotypes, is likely to be of someone who is relatively young, white, able-bodied, neither too fat nor too thin, neither too short nor too tall, and of average physical attractiveness. Thus, our studies of the stereotypes that accompany gender are most likely limited by their reliance on generalizations that systematically exclude such groups as old people, blacks, Chicanos, native peoples, Asians, disabled people, fat people, lesbians and gay men, and people whose appearance diverges markedly from the norm. It is also likely that, in holding to gender stereotypes based on the so called typical man or woman, our society creates added difficulties for individuals already victimized by racism, ageism, an intolerance of difference, or society's obsession with attractiveness and thinness.

Gender Stereotypes and Race

Researchers have tried to understand the interactive workings of racism and sexism by comparing the experiences of different gender-race groups. It has been noted, for example, that the forms of racism directed at black women and black men may differ, and that sexism may be expressed toward and experienced by a woman differently as a function of her race (Smith & Stewart, 1983). An examination of gender stereotypes cannot ignore the way race and gender stereotypes interact, or the ways in which different racial groups may differ in the images and ideals they hold for femininity and masculinity.

Femininity. In an attempt to determine whether the often-reported stereotype of women is actually just a stereotype of middle-class white women, Landrine (1985) asked undergraduates to describe the stereotype of each of four groups of women—black and white middle-class women, and black and white lower-class women—based on a list of 23 adjectives. She found that although the stereotypes differed significantly by both race and social class, with white women and middle-class women being described in ways most similar to the dominant culture's traditional stereotypes of women, all four groups were rated in ways consistent with this feminine stereotype. White women were described more often than black women by the most stereotypical terms: *dependent, emotional,* and *passive.* However, both groups were rated similarly on many other

From *Sex and Gender, An Introduction* by Hilary M. Lips. Copyright © 1997 by Hilary M. Lips. Reprinted by permission of Mayfield Publishing Company.

adjectives: *ambitious, competent, intelligent, self-confident,* and *hostile,* for example. The findings suggest that race and social class are implicit in what have been described as gender stereotypes, but that, to some extent, there is a set of expectations for women that transcends these variables.

According to the stereotypes held by the dominant white society, black women may be judged as less "feminine" than white women. The history of black women in the United Sates makes it clear that they have not, as a group, been conditioned toward the helpless, delicate, even frail image of femininity so often cultivated in their white counterparts. As slaves, women were expected to endure backbreaking physical labor and hardships, and were denied the security in their marital lives that white women took for granted. Angela Davis (1981) notes that, under slavery, black women constructed a definition of womanhood that included "hard work, perseverance and self-reliance, a legacy of tenacity, resistance and an insistence on sexual equality" (p. 29). According to some black women theorists, another legacy among American black women that sets them apart from white models of femininity is their strong tradition of speech rather than silence, of oratory, preaching, and storytelling rather than subservient listening (Collins, 1990). One view within the black community is that "blacks have withstood the long line of abuses perpetrated against them ever since their arrival in this country mainly because of the black woman's fortitude, inner wisdom, and sheer ability to survive" (Terrelonge, 1989, p. 556). Indeed, there appears to be a persistent stereotype of black women in the social science literature that encompasses strength, self-reliance, and a strong achievement orientation (Epstein, 1973; Fleming, 1983a).

Fleming (1983b) documented the existence of a "black matriarchy" theory among social scientists suggesting that black women are more dominant, assertive, and self-reliant than black males. While such a stereotype may appear positive at first glance, the narrow labeling of black women's strengths as "matriarchal" represents a refusal to conceptualize strong women in anything but a family context. Moreover, there has been a tendency to blame black women's strong, assertive behavior for some of the problems experienced by black men—a tendency that surely reflects stereotypic notions about what kinds of behavior are "proper" for women. Fleming and others (e.g., Dugger, 1991; Staples, 1978) argue that the black matriarchy theory is a myth based largely on flawed or misleading evidence. Fleming suggests that this particular stereotype may have arisen from the fact that "their long history of instrumentality in the service of family functioning may well have built in black women an air of self-reliance that arouses further stereotyping among those (largely white) social scientists more accustomed to white traditional norms for women" (Fleming, 1983a, p. 43). Indeed, Collins (1990) argues that the image of the black matriarch has been and is held up to all women as a symbol of the problems associated with a rejection of "appropriate" gender norms: "Aggressive, assertive women are penalized—they are abandoned by their men, end up impoverished, and are stigmatized as being unfeminine" (p. 75).

In contrast to black women, Chicanas have traditionally faced an ideal of femininity that glorified motherhood, subservience, and long-suffering endurance (Garcia, 1991). A woman was supposed to gain fulfillment by loving and supporting her husband and nurturing her children, and her place was unquestionably in the home (Astrachan, 1989). Historically, this ideal was fostered by the Catholic church, which played a large role in the colonization of what is now Mexico and the American Southwest. The church taught women to venerate the Virgin Mary and to emulate her by being silent, submissive, altruistic, and self-denying (Almquist, 1989). In recent years, however, Mexican-American women have taken on new roles. Their rates of participation in the labor force now resemble those of Anglos. In addition, a vocal Chicana feminist movement has emerged (Garcia, 1991; Segura & Pesquera, 1995). This movement, reflecting the values of Mexican-American culture, has incorporated a positive valuing rather than a rejection of the roles of wife and mother along with the acceptance of broader roles for women.

One of the reasons for race-related differences in gender-role attitudes is that members of different racial groups frequently live in different social conditions. For example, in the United States, black women have been more likely than white women to have to work for economic reasons; thus, working for wages has been a normative aspect of womanhood. For white women, however (and perhaps even more for Chicana women, who have only recently joined the labor force in such large numbers), the movement into paid work is more likely to provoke a questioning of traditional views of gender, since it represents a break with past traditions (Dugger, 1991).

Masculinity. The interaction of masculinity stereotypes with race is similarly complex. Black, white, and Chicano males in the United States, for example, hold and face differing, though overlapping, images of what it means to be masculine. The differences probably stem as much or more from variations in the social structural conditions currently facing the three groups than from their differing cultural heritage (Baca Zinn, 1989).

For example, the status of black males in American society has been in flux for the last several decades. Long after legal racial segregation ended, the effects of racism were still extremely obvious in the white society's treatment of black men. Black men were, in fact, often referred to *not* as men, but as "boys." Some writers have argued that it was not until the advent of the Black Power movement in the late 1960s that adult black males were recognized as *men* by most of American society (Franklin, 1984; Poussaint, 1982). As a group, black males currently live a hazardous existence as they try to maintain the competitive, success-oriented, aggressive aspects of masculinity in the face of discrimination and a lack of economic opportunities to pursue the "protector/provider" role. The result, according to some authors, is the channeling of attempts to be masculine into a cluster of behaviors that emphasize physical toughness, violence, and risk-taking, especially among *young* black men (Franklin, 1984). A tragic result is that the death rate from violence is frighteningly high in this group. Homicide is the leading cause of death among young American black men.

Richard Majors (1989) writes of the "Cool Pose"—a cultural signature adopted by many black males as a mechanism for survival and social competence. A set of poses and postures displays elements of control, toughness, and detachment that symbolize strength, pride, and a refusal to show hurt or weakness—especially to outsiders. According to Majors, "coolness" is also displayed through the "expressive life style." The style is manifested in creative, even flamboyant, performances in arenas ranging from the street to the basketball court to the pulpit: performances that say "You can't copy me." The Cool Pose, an aggressive assertion of masculinity, guards these males' pride and dignity, but may be costly to relationships when it blocks the expression of emotions or needs.

Chicano men have also developed a tradition that places strong importance on masculinity—expressed in an ideology of male dominance often labeled "machismo." Although some have argued that machismo is a myth used to negatively stereotype Chicanos and rationalize their lack of success in the dominant society, there seems little doubt that a male-dominant ideology of masculinity has played an important role in relations between Chicanos and Chicanas (Garcia, 1991). The core aspect of machismo is men's dominance over women. The stereotypical macho male expects to be considered the head of his household and demands deference and respect from his wife and children. He may feel strongly that his wife's place is in the home and that his manhood is demonstrated by the number of children he has (Astrachan, 1989). A Chicano man may behave in an aggressive, macho way partly as "a conscious rejection of the dominant society's definition of Mexicans as passive, lazy, and indifferent" (Baca Zinn, 1989, p. 95). Yet machismo appears to be, in its own way, a public pose. Researchers find that actual decision making in Chicano families is as likely to be egalitarian as male-dominant and that Chicano fathers are nurturant and expressive rather than distant and authoritarian with their young children. Even in egalitarian families, however, patriarchal ideology is strong, as expressed in statements that the father is the head of the house, the boss, the one in charge (Baca Zinn, 1989). Baca Zinn suggests that patriarchal ideology is associated with family solidarity and that perhaps "the father's authority is strongly upheld because family solidarity is important in a society that excludes and subordinates Chicanos" (p. 95).

It is possible that in a society that values masculinity over femininity, gender may take on a special significance for men of color and other men who have little access to socially valued roles. They may emphasize and act out the toughness and dominance aspects of masculinity, both as a defense against putdowns from the larger society and because "being male is one sure way to acquire status when other roles are systematically denied" (Baca Zinn, 1989, p. 94).

Citations

Astrachan, A. "Dividing Lines." In M.S. Kimmel and M.A. Messner, eds. *Mens Lives.* New York: Macmillan, 1989.

Baca Zinn, M. "Chicano Men and Masculinity." In M.S. Kimmel and M.A. Messner, eds. *Mens Lives*. New York: Macmillan, 1989.

Collins, P.H. *Black Feminist Thought: Knowledge, Consciousness, and the Politics of Empowerment*. Boston: Unwin Hyman, 1990.

Davis, A. *Women, Race, and Class*. New York: Random House, 1981.

Dugger, K. "Social Location and Gender-Role Attitudes: A Comparison of Black and White Women." In J. Lorber & S.A. Farrell, eds. *The Social Construction of Gender*. Newbury Park, CA: Sage, 1991.

Epstein, C. "Positive Effects of Multiple Negatives: Explaining the Success of Black Professional Women." In J. Huber, ed. *Changing Women in a Changing Society*. Chicago: University of Chicago Press, 1973.

Fleming, J. "Black Women in Black and White College Environments: The Making of a Matriarch." *Journal of Social Issues*, 39, 3 (1983): 41-54.

Franklin, C.W. II *The Changing Definition of Masculinity*. New York: Plenum, 1984.

Garcia, A.M. "The Development of Chicano Feminist Discourse." In J. Lorber & S.A. Farrell, eds. *The Social Construction of Gender*. Newbury Park, CA: Sage, 1991.

Landrine H. "Race x Class Stereotypes of Women." *Sex Roles*, 13, 1/2, (1985): 65-75.

Majors, R. "Cool Pose: The Proud Signature of Black Survival." In M.S. Kimmel & M.A. Messner, eds. *Men's Lives*. New York: Macmillan, 1989.

Poussaint, A. F. "What Every Black Woman Should Know About Black Men." *Ebony* (August 1982): 36-40.

Segura, D.A. & Pesquera, B. M. "Chicana Feminisms: The Political Context and Contemporary Expressions." In J. Freemen, ed. *Women: A Feminist Perspective* (5th ed.) Mountain View, CA: Mayfield, 1995.

Smith, A. & A.J. Stewart. "Approaches to Studying Racism and Sexism in Black Women's Lives." *Journal of Social Issues*, 39, 3 (1983): 1-15.

Terrelonge, P. "Feminist Consciousness and Black Women." In J. Freeman, ed. *Women: A Feminist Perspective*, 4th ed. Mountain View, CA: Mayfield, 1989.

Reflections

1. What are the gender stereotypes for gender? How do your observations or experiences affect the way you view this?

2. Which of your own stereotypes would you most like to change? Why?

6

While attention has usually been focused on the emotional costs of traditional female gender roles, the costs of traditional male gender roles have largely been ignored. According to Kimbrell, men can no longer afford to lose themselves in isolation, alienation from the family, and denial.

The Male Mystique

By Andrew Kimbrell

Men are hurting—badly. Despite rumors to the contrary, men as a gender are being devastated physically and psychically by our socioeconomic system. As American society continues to empower a small percentage of men and a smaller but increasing percentage of women, it is causing significant confusion and anguish for the majority of men.

In recent years, there have been many impressive analyses documenting the exploitation of women in our culture. Unfortunately, little attention has been given to the massive disruption and destruction that our economic and political institutions have wrought on men. Far too often, men as a gender have been thought of as synonymous with the power elite.

But thinking on this subject is beginning to change. Over the last decade, men have begun to realize that we cannot properly relate to one another, or understand how some of us in turn exploit others, until we have begun to appreciate the extent and nature of our dispossessed predicament. In a variety of ways, men across the country are beginning to mourn their losses and seek solutions. This new sense of loss among men comes from the deterioration of men's traditional roles as protectors of family and the earth (although not the sole protectors). And much of this mourning also focuses on how men's energy is often channeled in the direction of destruction—both of the earth and its inhabitants. The mission of many men today—both those involved in the men's movement and others outside it—is to find new ways that allow men to celebrate their generative potential and reverse the cycle of destruction that characterizes men's collective behavior today. These calls to action are not abstract or hypothetical. The oppression of men, especially in the last decades, can be easily seen in a disturbing upward spiral of male self-destruction, addiction, hopelessness and homelessness.

While suicide rates for women have been stable over the last 20 years, among men—especially white male teen-agers—they have increased rapidly. Male teen-agers are five times more likely to take their own lives than females. Overall men are committing suicide at four times the rate of women. America's young men are also being ravaged by alcohol and drug abuse. Men between the ages of 18 and 29 suffer alcohol dependency at three times the rate of women of the same age group. More than two-thirds of all alcoholics are men, and 50 percent more men are regular users of illicit drugs than women. Men account for more than 90 percent of arrests for alcohol and drug abuse violations.

A sense of hopelessness among America's young men is not surprising. Real wages for men under 25 have actually declined over the last 20 years, and 60 percent of all high school dropouts are males. These statistics account in part for the increasing rate of unemployment among men and for the fact that more than 80 percent of America's homeless are men.

The stress on men is taking its toll. Men's life expectancy is 10 percent shorter than women's, and the incidence of stress-related illness such as heart disease and certain cancers remains inordinately high among men.

And the situation for minority men is even worse.

Men are also a large part of the growing crisis in the American family. Studies report that parents today spend 40 percent less time with their children than did parents in 1965, and men are increasingly isolated from their families by the pressures of work and the circumstances of divorce.

The current crisis for men, which goes far beyond statistics, is nothing new. We have faced a legacy of loss, especially since the start of the mechanical age. From the Enclosure Acts, which forced families off the land in Tudor England, to the ongoing destruction of indigenous communities throughout the Third World, the demands of the industrial era have forced men off the land, out of the family and community, and into the factory and office. The male as steward of family and soil, craftsman, woodsman, native hunter, and fisherman has all but vanished.

As men became the primary cog in industrial production, they lost touch with the earth and the parts of themselves that needed the earth to survive. Men by the millions—who long prided themselves on their husbandry of family, community and land—were forced into a system whose ultimate goal was to turn one man against another in the competitive "jungle" of industrialized society.

The factory wrenched the father from the home, and he often became a virtual nonentity in the household. By separating a man's work from his family, industrial society caused the permanent alienation of father from son. Even when the modern father returns to the house, he is often too tired and too irritable from the tensions and tedium of work in the factory or corporation to pay close attention to his children.

While the loss of fathers is now beginning to be discussed, men have yet to fully come to terms with the terrible loss of sons during the mechanized wars of this century. World War I, World War II, Korea and Vietnam were what the poet Robert Graves called "holocausts of young men." In the battlefields of this century, hundreds of millions of men were killed or injured. In World Wars I and II—in which more than 100 million soldiers were casualties—most of the victims were teen-age boys, the average age being 18.5 years.

Instead of grieving over and acting on our loss of independence and generativity, modern men have often engaged in denial—a denial that is linked to the existence of a "male mystique." This defective mythology of the modern age has created a "new man." The male mystique recasts what anthropologists have identified as the traditional male role throughout history—a man, whether hunter-gatherer or farmer, who is steeped in a creative and sustaining relationship with his extended family and the earth household. In the place of this long-enduring, rooted masculine role, the male mystique has fostered a new image of men: autonomous, efficient, intensely self-interested, and disconnected from community and the earth.

Ironically, men's own sense of loss has fed the male mystique. As men become more and more powerless in their own lives, they are given more and more media images of excessive, caricatured masculinity with which to identify.

The primary symbols of the male mystique are almost never caring fathers, stewards of the land or community organizers. Instead, over several decades these aggressively masculine figures have evolved from the Western independent man (John Wayne, Gary Cooper) to the blue-collar macho man (Sly Stallone and Robert DeNiro) and finally to a variety of military and police figures concluding with the violent revelry of "Robocop."

Modern men are entranced by this simulated masculinity—they experience danger, independence, success, sexuality, idealism and adventure as voyeurs. Meanwhile, in real life most men lead powerless, subservient lives in the factory or office—frightened of losing their jobs, mortgaged to the gills and still feeling responsible for supporting their families. Their lauded independence disappears the minute they report for work. The disparity between their real lives and the macho images of masculinity perpetrated by the media confuses and confounds many men.

Men can no longer afford to lose themselves in denial. We need to experience grief and anger over our losses and not buy into the pseudo-male stereotypes propagated by the male mystique. We are not what we are told we are.

The current generation of men faces a unique moment in history. Will we choose to remain subservient tools of social and environmental destruction or to fight for rediscovery of the male as a full partner and participant in family, community and the earth?

There is a world to gain. The male mystique, in which many of today's men are trapped, is threatening the family and planet with irreversible destruction.

A men's movement based on the recovery of masculinity could renew much of the world we have lost. By changing types of work and work hours, we could break our subordination to corporate managers and return much of our work and lives to the household. We could once again be teaching, nurturing presences to our children. By devoting ourselves to meaningful work with appropriate technology, we could recover independence in our work and our spirit. By caring for each other, we could recover the dignity of our gender and heal the wounds of addiction and self-destruction. By becoming husbands to the earth, we could protect the wild and recover our creative connections with the forces and rhythms of nature.

Ultimately we must help fashion a world without the daily frustration and sorrow of having to view each other as a collection of competitors instead of a community of friends.

Shortly after World War I, Ford Madox Ford, one of this century's greatest writers, depicted 20th century men as continually pinned down to their trenches, unable to stand up for fear of annihilation. As the century closes, men remain pinned down by an economic and political system that daily forces millions of us into meaningless work, powerless lives, and self-destruction. The time has come for men to stand up.

Reflections

1. What are the costs of being male in our society?

2. Who, in your opinion, contributes to the perpetuation of traditional male gender roles? Do you agree with these roles? Why or why not?

3. How have traditional male gender roles affected you?

The author is one of the leading researchers on love in this country. In this excerpt from his book, Sternberg combines the results of social science research on love with a practical application. As you read, try and reflect upon your own experiences of love and compare them to the characteristics described below.

Making Love Work

Excerpted from *Love the Way You Want It*
By Robert Sternberg

You Tolerate Ambiguity Relationships are inherently ambiguous. Much of the time, it is hard to pinpoint exactly what is going on. Some people are willing to accept ambiguity as a given in a relationship and work to improve communication to reduce its negative consequences. But for others, ambiguity is a constant source of frustration and unhappiness.

In the first blush of love, we make promises, and mean them. We say, "I'll never leave you," "I'll never hurt you," "I'll always feel this way about you." We try, with our words and our wills, to pin down the future, in the same way that the Three Little Pigs sealed their house against the wolf's attack. But there is no way to be absolutely secure in life. People who are successful at love accept ambiguity. More than that, they are able to rejoice in the mysterious turns of life.

You See Obstacles as Challenges Sooner or later, every relationship runs into obstacles. They might be financial, parental, career-related, sexual, or any number of things. People who thrive in relationships are those who are willing to accept obstacles as challenges rather than signs of defeat.

Unless you accept obstacles as part of the challenge of a relationship, it is unlikely that you will be very successful. Moreover, your ability to overcome difficulties will be greatly enhanced if your partner feels the same way. It is much easier to get over the rough spots if both of you strive to do so, rather than if one of you carries all the burden. If one of you carries the burden, you may force change occasionally, but it probably won't work over the long term. When you both contribute, you can usually overcome your problems. Indeed, this is one of the cornerstones of a good relationship.

Viewing your problems as challenges is not a stoic posture. It doesn't mean "grin and bear it." Rather, it is having a deep curiosity about life, a tendency to ask, when problems arise, "What is this teaching me? How can I turn this problem into something of value?"

You Embrace the Future "Nothing ventured, nothing gained," goes the old saying. People who succeed in relationships are those who are willing to try new things. They are not afraid to open themselves up to the possibilities that the future offers. If you remain stuck in the present, with an eye to the past, you may never get hurt—at least, not in the way you have learned to think about hurt. But you'll never be where the action in life is, either.

Embracing the future does not mean that you behave like a stuntman, crashing against life at every turn. Being a thrill-seeker won't help you achieve a mastery of love. The challenge is not to seek out risks, but rather to accept them when they present themselves.

You Seek Growth Truly creative people are always growing. They understand that life is a dynamic process and that to stand still is to stagnate. The myth of living "happily ever after" denies the necessity of growth and change. Life is not like that, and neither are relationships. Relationships require ongoing renewal, and sometimes they change in major ways. Successful couples are not afraid to change and grow.

There is always the danger in a relationship that you might grow apart. When you first meet someone, you may find that you are compatible and that you get along well because you are at a point in your lives when your needs are closely matched. But it's possible that the person who attracts you now may not have attracted you at another time in your life when your mutual needs were not as similar. It is important to realize that the closeness you feel now may not continue or be easily maintained. The most difficult task in a relationship is often being able to grow together rather than apart. To accomplish this, you need to put as much communication, support, and sharing into your relationship as possible.

The willingness to grow and take responsibility for one's life is a fundamental quality of a successful relationship. Furthermore, this quality is essential to the ability to live productively. When we allow ourselves to be stuck in a negative self-story, we do not take advantage of the gift of life.

You Believe in Yourself There will be many times, during the course of a relationship, when you will find yourself questioning your judgment, possibly even your sanity. There may be times when you begin to lose belief in yourself and begin to doubt that you have made the right decisions. Indeed, it is impossible to be involved in a close relationship without making mistakes. But one of the most important lessons about intelligence is that smart people are not people who never make mistakes; they are people who learn something of value from their mistakes. The ultimate test of Relationship Intelligence is how much you learn from your mistakes and how well you apply what you learn.

An important skill to develop is the ability to detach yourself from your mistakes and view them objectively. Ultimately, you are the person you are, not only because of your success, but also because of your failures. But whether you succeed or fail, you need to keep the courage of your convictions and your belief in yourself. If you don't believe in yourself, you can't expect anyone else to believe in you.

You Are Willing to Forgive I know people who pride themselves on never forgetting or forgiving a slight. They possess photographic memories for what they perceive to be insults or wrongs, and they are willing to wait months, sometimes years, to get their revenge. Or they "get back" in more passive ways, reminding the person who hurt them of the past behavior. As a result, they continue, year after year, to throw hand grenades into the relationship.

If you find that you are struggling to forgive your partner for a past mistake, you should examine what your end goal is. Are you interested in finding a way to make your relationship better, or are you more interested in punishing your partner and making him or her suffer? Your answer to this question will determine whether you will be open to resolving your conflicts and growing in your relationship.

You Can Accept Others as They Are Most people have at least some desire to exert control over others. We imagine what we would like other people to be, and then, consciously or unconsciously, we try to mold them into the image we have created. Some of the changes for which we work may be genuinely constructive. But too often we ask others to be what they are not. Sometimes people fall in love with an image, not the reality, and spend their relationships resenting that their partners cannot be what they never were.

You are never going to find anyone who is the perfect image of what you are looking for. So find someone whose perceived flaws are ones you can live with and accept them, just as you hope that person will accept your flaws. This is not to say that you should never try to help a person improve. But you must understand the improvement can't be forced from the outside; it must come from within.

Sometimes the thrill or romance can dazzle us to the point where we put off confronting the truth about a person. We avoid examining real issues, such as: Is this a person whose goals and life-style match what we want in life? A decision to commit to another person is also a decision to accept that person for what he or she is.

You Are Optimistic A bottle half empty is also a bottle half full. How you describe it has to do with your inner sense of optimism. It is a lot easier, and generally more satisfying, to live with a person who sees the bottle half full than one who sees it half empty. Almost any turn of events can be construed in a way that is either more positive or more negative. If you turn every disagreement into a major crisis and every misfortune into a major disaster, or if you look for the downside of every happy event, you will succeed not only in making yourself miserable, but in doing the same for those around you.

Some people wait for happiness to come to them, expecting it to appear as a bolt out of the blue. They do not seek it out, nor do they recognize it in the life they have. If you do not possess the quality of optimism, it will be hard for you to truly succeed in a relationship. Relationships constantly demand that we operate from our hopes not our fears. They challenge us to find the possibilities in every struggle. If you're predisposed to look at the gloomy side, even the everyday hassles of being in a relationship might feel overwhelming.

You Have Patience When things aren't going our way, we often need patience to allow time for them to work themselves out. We do not live in an age when people are very good at delaying gratification. But sometimes we would be much happier if we were willing to wait for gratification, rather than insisting on having everything right away.

In matters of love, we are especially reluctant to let time take its course. We cannot stand to be without a mate and when we find someone we like, we want to rush into commitment. There is a sense of desperation underlying our search for love, a fear that we might miss out on it altogether and be lonely forever.

Successful relationships are not born from desperation. Nor do they happen overnight. It takes time to develop the bond of intimacy that is needed to make a relationship strong and enduring. The term "love at first sight" would be better phrased as "passion at first sight." And the urgency of passion is not a sufficient foundation for a lasting relationship.

Your Love Is Selfless If I were asked the single most frequent cause of the destruction of relationships, and the single biggest contributor to "dumb love," I would say it is selfishness. We live in an age of narcissism and many people have never learned or have forgotten how to listen to the needs of others. The truth is, if you want to make just one change in yourself that will improve your relationship—literally, overnight—it would be to put your partner's interests on equal footing with your own. Selflessness does not mean that you sacrifice your own wants and needs for another person's. Rather, it is the ability to achieve a balance in your relationship. You respect your partner and consider his or her needs as being just as important as yours. And you possess the quality of compassion—the ability to recognize and respond during those times when your partner needs more attention than you do.

Selfishness is usually based in a fear people have that, once they open the door, they will be swept away by a violent tide and soon lose all of their freedom and individuality. On the other hand, I have observed people who *always* put others' needs ahead of their own, claiming that their needs are not worthy of equal consideration. The concept of equality transcends both of these responses. The quality of selflessness allows two people to be everything they can be as individuals, while they grow as a couple.

Reflections

1. Do you think that the qualities described in this article are the most important ones in a love relationship? Why or why not?

2. Which qualities would you exclude? Include?

3. Do you see any qualities in current or past relationships which need work?

4. Would you be willing to share your reflections on this article with a partner?

8

An expert and voracious writer in the field of love, Leo Buscaglia draws from his teachings, research and insights to unveil and bring to light the topic of jealousy. In this excerpt, he discusses responsibility, acceptance of one's limitations, and the freedom that comes when one lets go.

On Jealousy

Excerpted from *Loving Each Other*
By Leo Buscaglia

Since few of us who choose to form relationships with others will be totally free of jealousy, perhaps it would be best to look at better, more lasting ways to come to terms with the emotion. The great psychoanalyst/philosopher Theodore Reich has said, "Jealousy is a sign that something is wrong, not necessarily rotten, in the organism of love." Perhaps seeing jealousy as a warning of "something wrong" is the first positive step to its being corrected, since to fight or try to negate jealousy doesn't actually solve anything. The only real solution to jealousy seems to be to work it out. Feeling a strong emotion is necessary to making changes. Anthropologist Margaret Mead has suggested that jealousy is an emotion which is "a festering spot in every personality so affected, an ineffective, negativistic attitude which is more likely to lose than gain any goal." But, she admits that it may be of value for it may be responsible for the passion, the intensity, from which is born enterprise. In her anthropological studies of the people of Samoa she found no jealousy, but she also found few strong feelings, competition, or motivation.

We are responsible for our jealousy, no one else. Blaming others for what we feel, can lead nowhere. Change will only begin when we are willing to accept our jealousy as our responsibility, not necessarily bad unless negatively acted upon. Relieving others of our responsibility, we can then begin the productive processes necessary for finding out what can be done about it.

Rollo May, the famous analyst, has said: Jealousy requires turning one's attention to one's self and asking why is my self-esteem so low in the first place? I quite understand that this question may be difficult to answer. But at least it turns your concern to an area you can do something about.

Persons who cling to jealousy destroy *themselves*. They use energies for dead-end feelings which could be channeled into creative solutions. Of course, no one chooses to be jealous, it simply happens. What is essential is to change the values and beliefs which created the response. Jealousy generates much feeling, but actually produces little action. It becomes an insidious process which keeps us from seeing accurately what there is. It nurtures only itself. It succeeds only in making us feel impotent. As such, jealousy is most often a product of our personal insecurity and low self-esteem. It occurs because we see ourselves as having less to give than the object of our jealousy. It steals our rationality. We become unable to see our strengths and allow ourselves to be overcome by what we are convinced are weaknesses and inadequacies. We feel valueless. We lose our sense of dignity and worth. We become frenzied, paralyzed or afraid to act. We forget the simple fact that because someone does not elect to meet whatever conditions have been imposed in our relationship, our true inner value as a person is not diminished, nor is theirs. We forget that we cannot force anyone to meet our needs, to be what we want them to be, do what we want them to do, respond as we would have them respond or feel what we think they should feel. This is a human impossibility, an illusion, a fantasy. Even if the other person concedes to being "ours," at best that is only a figure of speech.

© 1984 by Leo F. Buscaglia, Inc.

Perhaps we must finally accept the fact that we can never possess another human being. A decision to unite is an agreement between two separate units, which will always, in a sense, be separate. We must learn that loving others is to want them to be themselves—painful as it may be—with or without you. After all is said and done, what else can we do but wish them well? If a friend or loved one wants to go, even if we devise a hundred ways to try to hold on, we will never be successful. And how little we value ourselves when we manipulate someone in order to keep them, when they would rather be elsewhere. We are better off without those individuals in our lives.

Jealousy diminishes only when we regain a feeling of worth and self-respect, stop internalizing the problem and begin to view it objectively as something stemming from our personal demands and needs. These may arise from our desire for status or loyalty. They may be due to our insecurity, our need to control, to possess, our need for exclusivity, or fear of loss of face.

Loyalty in a relationship is based upon trust and respect. It can only be offered, never demanded. It is based upon voluntary devotion. Relationships are continually changing. A mutual agreement to be loyal or honest will form the basis from which future trust will arise. Loyalty is, therefore, a pact. Fidelity is a pact. The earlier these qualities are discussed and agreed upon in a relationship, the more secure the future of the relationship. Of course, the decision must be mutual, and the decision must be binding. Any change in the expectations over time must be accepted, discussed, and new decisions formed.

The word "jealousy" stems from the Greek work "jeal." It suggests that a valued possession is in danger and that some action must be taken. It implies that what can be seen as a negative phenomenon can be changed to a positive one over time. As the relationship, and those involved in it, becomes stronger and more secure, so will jealousy be minimized. Learning to let go, since most of us believe that love is based on "holding on to," is very difficult. Perhaps the greatest love presupposes the greatest freedom. There is an old saying which suggests that love must be set free, and when it comes back to you, only then will you know real love.

When we have finally conquered extremes of jealousy, we will emerge better and stronger lovers. We will understand the joy and strength which comes from solving our own problems, meeting our own needs, and loving freely without demands. As always after having conquered something of a lower order, we will be lifted to new and greater heights.

Eleanor Roosevelt said:
Every time you meet a situation, although you think at the time it is an impossibility and you go through the tortures of the damned, once you have met it and lived through it, you find that forever after you are freer than you were before.

Don't be afraid of jealousy. It is a natural and normal emotion. Everyone who cares and loves feels jealous at one time or another. The essential decision is whether you will allow your jealousy to become an all-consuming monster, capable of destroying you and those you love, or become a challenge for you to grow in self-respect and personal knowledge. The challenge will rest with you.

Reflections

1. Do you hold any feelings of jealousy?

2. From where do these feelings come?

3. How might you begin to overcome feelings of jealousy?

4. Do you think it's possible or desirable to eliminate jealousy?

9

In this reading, the noted author of books for both children and adults, Judith Viorst, suggests fifteen ways to keep fights fair. By fighting fairly, couples can use their inevitable conflicts to solidify rather than destroy their relationships.

How to Fight Fair

By Judith Viorst

I was not fighting fair when I accused Milton of refusing to buy a new car because he was saving money for the teenaged nymph he was doubtless planning to marry the minute I died an untimely death after the engine of our miserable blue station wagon stalled (yet again) in the middle of a highly trafficked intersection.

Wally was not fighting fair when he blamed his tension and subsequent defeat in the Shadybrook Tennis Club tennis finals on Annabel's not wanting to make love that morning.

And Shari was not fighting fair when she told Mike, who had promised to show up at three and showed up at four, that he was unconsciously perceiving her as the feared and hated mother of his childhood and making her pay for his mother's rejection of him.

No fair!

Now you may want to raise the question of whether it's possible in a marriage to feel furious, hurt, humiliated and/or betrayed and still have sufficient self-restraint to hit above the belt, to fight fair. And you may want to ask another, far more fundamental question: Can two normal people be married and never fight?

I'll start with the second question, to which I have a simple answer: Certainly not.

For the first thing we need to concede is that marriage is a difficult living arrangement and that, as William Dean Howells once wrote, "...the silken texture of the marriage tie bears a daily strain of wrong and insult to which no other human relation can be subjected without lesion."

And the next thing we need to concede is that, much as we deeply love and respect our spouses, there will be times when we deeply hate and despise them.

And the third thing we need to concede is that, rational and civilized though we may be, not all our marital differences can be resolved by simply sitting down and talking about them. Sooner or later, some feelings we feel and points we need to make will have to be communicated more...vigorously.

That vigorous form of communication is what most people mean when they use the word "fight."

Actually, I have become considerably less of a marital fighter in recent years—I just don't enjoy a good fight the way I used to. And although I still firmly believe that it is sometimes absolutely essential to have a fight, I am now more inclined to ask myself before I launch an attack—or before I respond to one—"Is this fight necessary?"

By that I mean: Will this fight resolve a current problem or prevent a future problem or provide a badly needed emotional outlet? Or will we both just wind up feeling bruised? And sometimes—not always, but sometimes—having decided a fight can't do anything but hurt, I am able to persuade myself not to fight.

But what is a wife who decides not to fight to do with her feelings of anger? Linda offers one ingenious solution. When her husband Phil called from Albany and said that he was tied up and couldn't make the party that he'd practically sworn on his life he was going to make, Linda was apoplectic, but she also knew that she couldn't change his mind. "Hang on, Phil," she told him. "Let me take this in the kitchen." And she put down the phone and ran from the second floor, screaming at the top of her lungs all the way down the stairs: "Miserable, rotten, no-good son of a..." and a few other words I'd just as soon not mention. After that, she said, she was able to lift the receiver and ask, with relative calm, "So tell me, Phil, what time *should* I expect you?"

Now there is no point in not having a fight if what we have instead is a cold war (no talking except for basic queries like "Would you consider it a vicious attack on your masculinity if I asked you to pass the butter?"), or if we translate our repressed rage into bleeding ulcers or headaches, or if we engage in some major marriage-threatening activity like taking a lover or hiring a divorce lawyer. But if we have to fight, and if we want to preserve—not massacre—our marriage, what are the limits we need to set to keep the fight we're fighting clean and constructive? Here, from some fighters I've talked with, are several suggestions.

1. *Avoid paranoid overstatements.* (See Milton and Judy's fight.)

2. *Accept responsibility for your own failures.* (See Wally and Annabel's fight.)

3. *Do not practice psychiatry without a license.* (See Shari and Mike's fight.)

4. *Don't wait too long.* We need to consider the story of the genie in the bottle who, during his first thousand years of incarceration, thinks, "Whoever lets me out will get three wishes," and who, during his second thousand years of incarceration, thinks, "Whoever lets me out I'm gonna kill." Many of us, like that genie, seem to get meaner and more dangerous the longer our grievances are bottled up.

5. *Know what you want.* My friend Nina says that her idea of a clean fight is the delivery of the following crystal-clear message: "I'm upset; here's why I am; here's what I want"—though it may take some time to figure out what she wants. Recently, for instance, while fighting with her husband yet again about working late at the office night after night, it suddenly struck her that she didn't actually want him to say to hell with the work and miss his deadline, that all she really wanted was for him to say, "I miss you. I miss the baby. I feel terrible about not being home. And you're such a fabulous person for being able to handle everything while I'm gone."

She got it.

6. *Figure out what you're really, really fighting about.* There are many battles about, say, his forgetting to make hotel reservations for the car trip out West, or his failing to fuss over you when you get sick, that blow up into something utterly out of proportion and nasty because what you're really fighting about, and what's getting you so upset, is the thought that he wouldn't have been (a) so negligent or (b) so uncompassionate IF HE LOVED YOU.

But wait a minute. You may be able to avoid hurling a lot of unpleasant accusations at him if you can recognize that what you are really, *really* fighting about, as is often the case, is differences in style, not lack of love. You may need to recognize, for instance, that although, while you were growing up, your daddy always made hotel reservations and got a million brochures and planned every step of a family trip in advance, some people like to improvise when they travel—that improvisation isn't the same as negligence. You may also need to recognize that although, while you were growing up, a sick person got ten glasses of orange juice and tons of attention and permission to fool around with Mommy's jewelry box, your husband's family was stoic about sickness—that not fussing doesn't have to mean not caring. And once you figure out that the real, real fight has nothing to do with IF HE LOVED YOU, you are more likely to have a clean and constructive fight.

7. *Stick to the point.* If you are having a fight about the way he is handling the children, I promise you that it will not advance your argument if you also note that he is overdrawn at the bank, talks with his mouth full and frequently leaves you the car with no gas in the tank.

8. *Stick to the present.* A sense of history is a wonderful thing. Total recall is certainly impressive. Memory and the long, long view surely contribute to the richness of life. But reaching back in time for crimes committed, say, yesterday, contributes—and I speak from experience—only trouble.

For several years, when fighting with my husband, I displayed my capacity for total recall, providing him with what I rather smugly like to call "a sense of perspective." I would, for instance, point out to him that not only had he forgotten to pick up my blouse at the cleaners' that afternoon, but that he had also failed in his stop-at-the-cleaners' assignments on the following 14 occasions. I would then list the dates and the garments he hadn't picked up, starting with the beige chiffon dress that—because of his carelessness—I wasn't able to wear to our engagement party.

After a couple of decades of this, however, it began to dawn on me that never once did Milton reply to my historical references with a "Hey, thanks for pointing out this destructive pattern of mine; I sure do appreciate it." Instead he replied with, "Bookkeeper! Scorekeeper!" and other far less charming epithets, and the fighting would deteriorate from there. I now follow—and strongly recommend—the statute of limitations my friends Hank and Gail have established: "No matter how perfectly something proves your point, you can't dredge it up if it's more than six months old."

9. *Never, never attack an Achilles' heel.* If your husband has confessed to you that his cruel high-school classmates nicknamed him "The Hairy Ape," and if, in adulthood, he still has fears about being furrier than most, you can—when you are furious—call him every name in the book, but you can't call him that one. I have a friend who (excuse the mixed metaphor) located her Achilles' heel in "my fat butt, which I always worry about and which my husband has always assured me he loves." She says that if, in the course of a fight, he told her that he'd never liked her butt, she might never forgive him.

Two Achilles' heels that are mentioned so often that they must be universal are sexual performance and parents. It seems that it is tricky enough, in life's mellowest moments, to discuss sexual dissatisfactions with a mate; but to scream in the heat of battle, as Louise confessed she once did, that "At least my first husband knew where to put it," is a rotten idea. And so is calling your mother-in-law an "old bat," even though her own son—your husband—has used those very words on many occasions. For some reason, we all seem to feel that although we're allowed to criticize our parents, it's dirty pool for our spouses to be doing it.

10. *Don't overstate your injuries.* Don't claim—either directly or indirectly—that he is giving you migraines or destroying you psychologically unless he is. Oh, I know that it is sometimes tempting to score a few points by pressing your fingers to your temples and saying, in the middle of a fight, "Quick, get me my pills, the pain is blinding," or "Maybe I ought to go into psychoanalysis." It is also sometimes tempting to burst into tears—particularly of the "you're so brutal you even make strong women weep," muffled-sob variety. But don't. For overstating your injuries is not just dirty fighting; it is, sooner than you can imagine, ineffective. If you want him to believe that he's gone *too* far when he's gone too far, you've got to try to maintain your credibility.

11. *Don't overstate your threats.* Many people, in the course of a fight, make threats they don't mean, like, "If you don't slow down, I'm getting a divorce." The trouble with these threats is that there is always the risk of being called on them and either actually having to, say, get out and walk or losing a lot of face. Like overstatements of injury, these threats—if not followed through—are subject to the "boy who cried wolf" syndrome. Even worse, I know two women whose "if you don't stop seeing her" ultimatums converted painful marital fights into fierce and fatal power struggles, ending in divorces that neither couple, I am convinced, actually wanted.

12. *Don't just talk—listen.* You needn't go along with a word he says, but shut up long enough to let him say it. And pay attention. For as a lawyer friend likes to point out to her spouse in the midst of a fight: "You can disagree with what I say as long as you can repeat my views to my satisfaction."

13. *Give respect to feelings as well as to facts.* You are not allowed to say, when your husband tells you that he is feeling ignored or put down, "That's absurd. You shouldn't feel that way." You may argue that perhaps he is overreacting or misinterpreting, but you have to acknowledge that he feels what he feels.

14. *There needn't be one winner and one loser.* You both could agree to compromise. You both could agree to try harder. You each could understand the other's point of view. You also could lose without treating it as a defeat. You also could win (this really takes class!) without treating it as a victory.

15. *When you're finished fighting, do not continue sniping.* This includes doing take-backs like "I said I'd go but I didn't say I'd enjoy it," or staring bitterly out a window because, as you explain when asked what's wrong, "I guess my wounds don't heal as quickly as yours do." Nor, after a fight, is it useful to murmur things like "Why is it always my fault?" or "Why am I the one who makes all the concessions?" or—as in the fight between Milton and me—"Now that you've agreed to buy a new car, I only hope I survive until it gets here."

Though of course we are bound to slip, we at least should try to follows these tips. Let's face it. All is *not* fair in love and war. Clean and constructive fighting is better than down and dirty fighting, that's for sure. And fighting by the rules could help us live happily—although scrappily—ever after.

Reflections

1. As you read through the author's fighting rules, which ones do you tend to follow and which ones do you tend to ignore?

2. Do you use different rules when you fight with different people, such as friends, dates, partners, siblings, parents, or children? Why? From whom did you learn these rules?

3. What other rules can you add to the author's list?

10

Secrets have been a staple of our American society; the most dangerous and most common operating within the family itself. In this reading, the authors explore some of the secrets and lies that haunt some families and the freedom that can occur when truth is revealed.

Family Secrets

By David Gelman and Debra Rosenberg

Knowledge is power in Washington, and none are more powerful than those who traffic in the secrets of state. For 25 years Madeleine Albright has lived in this milieu, moving confidently from the casual confidences of the legislative lobby to the pinnacle of diplomatic posts, where she is fed a daily diet of eyes-only, top-secret papers. But as the world now knows, even in a job that starts each day with spy-master reports and top-secret briefings, there can be another kind of powerful secret all too familiar to Albright's countrymen: secrets of the soul.

Thanks to the disclosures of the last fortnight, she is no longer—publicly or privately—merely the stunningly successful daughter of brave Czech immigrants. Now she is also the child who was kept in the dark of secrecy—never told that her grandparents were Jews who perished in the smoke of Auschwitz and Terezin; raised falsely on a glorious but invented history of Prague Easters and peaceful Catholic deaths. But for her position as secretary of state, Albright would be allowed to come to terms quietly with her new history. She is, after all, too accomplished, too mature, too busy, to have her identity reshaped at this late date. But public figures, in this age of full disclosure, don't have the luxury of private space and time. "I've seen the questions from people wondering why I didn't put it together," she said last week in an exclusive interview with NEWSWEEK. "I regret that deeply. This is not a good analogy, but let me say if it never occurred to you that you were adopted, why would you think that you're adopted? It was not a question."

Few revelations can be more unsettling than those that tamper with our own notions of identity. To learn—after years of carrying a sense of self as intimate as one's own skin—that you are not really the person you thought you were can be as shattering as going through an earthquake. Especially when you're young. "It unseats your faith in the order of the universe," says writer Letty Cottin Pogrebin, who fainted when she was abruptly told that her idealized parents had both been married before, and that her older sister was, in fact, her half sister. When she revived, she found her parents kneeling beside her, apologizing for not having told her earlier. She was 12 at the time. At 57, Pogrebin retains an unassuageable skepticism about surface appearances. "You can't take anything for granted," she says. Lesli LaRocco, who first learned at 27 that she was adopted, says the knowledge has shifted her whole perspective. "People will tell you it doesn't matter ... but in fact it does change everything," she says.

Secrets, personal and otherwise, have become common currency in an era that holds the sex life of presidents no more private than the mating habits of elephants. Like it or not, we now know the commander in chief's choice in underwear and that he may have a half sibling, left behind by his traveling-salesman father. Once scandalous matters like divorce, infidelity and illegitimacy have been devalued, their power to shame defanged. People now advertise their intimate problems on TV talk shows, or even on the Internet. Click on the "alt.support.step-parents" newsgroup and behold this posting, from Sandi: "I am a single parent of a two year old boy. His father left me a week after I told him I was pregnant ... My son is in daycare and he sees the other kids with Moms and Dads ... What do I tell him?"

As a society, we are neither as open-minded nor as shockproof as we affect to be. There is still a list of secrets families keep, but it has been abbreviated to such issues—criminal records, mental illness and homosexuality—which remains tightly closeted in many parts of the country. AIDS is often covered up. Suicide can still prompt families to close ranks against the curious.

Some secrets are pernicious, and the most damaging operate within the family itself. They create peculiar gaps in communication, a "zone of silence" in which certain matters cannot be discussed. Sometimes everyone knows the secret, says family therapist Evan Imber-Black, director of family and group studies at Albert Einstein College of Medicine in New York. "But the unspoken rule is, we're not allowed to know we know. So relationships get shot through with silences, sudden changes of subject. When there's a central secret, people start talking in funny ways, almost in code." They form alliances, dyads and triads, in which some are "in" on a secret and some are not. "More and more topics go off-limits," she says.

Secrets beget secrets; lies beget other lies to sustain them. Covering up for a drunken parent may plunge the family into a whole pattern of concealment. "You see it in families where there's substance abuse," says Imber-Black. "Where the rule against noticing things becomes the overarching way the family does business, that's going to play out around other things as well." Children have great radar for these blank, forbidden areas, and they can be especially vulnerable to the fallout. Often parents will come into therapy with the complaint that a child is lying. "Time after time," says Imber-Black, "we find lying is in the family; it's not something the child has invented ... When they are being tricked in some way, they will act out the same behavior, metaphorically."

Lies are thus transmitted across generations. In one typical case, for which Imber-Black functioned as a consultant to family therapist Peggy Papp, a mother brought in her 15-year-old son Kevin, who had been failing at school and stealing money from his parents. "The boy is lying," Imber-Black remarked to Papp after one therapy session. "Where is the lie in the system?" The "system" was eventually traced to Kevin's great-grandmother, whose penchant for lying and petty theft was never discussed in the family. Also, Kevin's mother herself had been perpetuating the cycle by hiding the truth about his father, an alcoholic and a drug dealer, because she was terrified the boy might follow in his footsteps. Papp first debunked the mother's notion that lying and stealing are genetic. Then she suggested to the mother that "Kevin was being loyal to the tradition of secrecy in the family by being secretive," and that the best treatment was the truth.

Papp's and Imber-Black's insights helped Kevin's mother break the tension between herself and her son by talking candidly about the father who had abandoned the family when Kevin was an infant. A year later the boy's misbehavior had all but ceased, and he was doing much better in school. "An adolescent's lying and stealing, or a young woman's bulimia, may be ways to comment metaphorically on the unmentionable," says Imber-Black. "A symptom may be a symbolic expression of powerful emotions connected to the secret."

Most families have secrets, and surprisingly, adoption continues to be one of the most common. Under enlightened practices, children are told about adoption more readily than in the past. But in the interest of protecting them, many parents keep adoptees in the dark, setting them up for possible future shock. When they eventually find out, there is, among the rush of sensations, a crushing sense of betrayal—of being the victim of a kind of elaborate hoax. Lesli LaRocco can laugh now when she recalls some of the ludicrous deceptions that were used to conceal her adoption. She doesn't know how she missed the clues—all the subjects never discussed, the questions not answered, from innocent ones like "Mommy, did I come from your tummy?" to more troubling ones like "Why am I so tall when everyone else is so short?" Her grandmother once offered a reason: "I have some very tall cousins." LaRocco, in fact, was a wispy blonde in a family of stocky, dark-complexioned Sicilians, but she never questioned the anomaly. "Children get signals from their parents about what's off-limits," she says.

LaRocco's parents were separated, and she didn't learn she was adopted until a few months after her mother died, when the man she had known as her father for the 27 years of her life told her. For a time, she couldn't trust anything. It wasn't only the one big lie she had been told, but "the 1,000 lies that supported it," she wrote in an adoption Web site. She began reading everything she could find about the subject and diverted her growing rage into anger at the adoption system itself. "It was easier to be angry at the system ...

than it was to be angry at my dead mother," she says.

LaRocco remains indignant at the system that perpetuates the still undisclosed secret in her life: the identity of her birthparents. "The state participates in the lie," she says. She has joined an adoptee activist group called Bastard Nation, which distributed leaflets at screenings of "Secrets & Lies" pointing out that, unlike Britain, 47 U.S. states still deny adoptees access to birth records. Most such domestic secrets are even crueler, because they are unnecessary, LaRocco believes. "In the end," she says, secrets "are always destructive, because they are based on shame."

It's the secrets born of shame, embarrassment or desperation that are most likely to surface sooner or later. "Most people can't keep that kind of secret," says Dr. Carol Nadelson, a psychiatrist at Harvard Medical School. "You're always in this danger zone. You can't have anybody know. You build one lie on another lie. You can't be free of it."

Some people nevertheless bring it off—while the secret is in their control. Madeleine Albright's father, Joseph Korbel, apparently never wavered in his assumed guise. Colleagues at the University of Denver, where he taught from 1949 until his death in 1977, were astonished to learn that he was Jewish. John Silber's story closely parallels Madeleine Albright's. Silber, a chancellor and former president of Boston University, was baptized in the Presbyterian church. He learned catechism, went to Sunday school, took his first communion and sang in the choir there. In the church also was a stained-glass window that his father had donated in honor of Silber's grandmother Elizabeth Silber. Then, at 33, while in Germany on a Fulbright scholarship, Silber met a cousin who expressed some surprise that he didn't consider himself Jewish, like some other members of the family.

Silber was stunned. His father's relatives, like his father, had never struck him as anything but observant Christians. But he later learned that one of his father's sisters had died in Auschwitz, and that his great-grandfather had been a renowned Jewish scholar and artist in Berlin. Silber's own father had come to the United States as a sculptor on the German pavilion at the 1904 World's Fair in St. Louis. He had never alluded to his Jewish background—not even to Silber's mother, who was equally shocked at the news. Silber himself has adjusted to the idea. "I am who I am, and that is not altered by this," he says. He believes his father freely chose to convert to Christianity and freely chose not to discuss his decision. "He slammed the door shut on that part of his background, and he kept the door shut," Silber says. Still, even at the age of 70 Silber feels a gnawing doubt. What troubles him is that while he had considered himself "a very close friend" of his father's, the disclosure seemed to call their whole relationship into question. "I realized I did not know my father as well as I thought I did," he says.

The stress of bottling up secrets takes a psychological toll. Deborah Blanchard has direct evidence of physical effects. In the late 1940s Blanchard, who is white, entered the New England Conservatory and fell in love with a black man. Both families objected, but the couple moved to a black neighborhood in Boston and had a son. By the time Blanchard was pregnant with their second child, the marriage was dissolving. She divorced her husband and moved back to her parents' home in Lowell, Mass. Her sons were not accepted in the white community there, and when the oldest reached school age, she decided the children would be better off living with a black family. She put them up for adoption, then tried to resume her life. But she was tormented by the decision and, oddly, she lost her trained lyric-soprano voice. "I was never able to sing after that," she says.

But that wasn't the end of the story. In 1963 Blanchard got married again, to a white man. She mentioned the earlier marriage only vaguely and said that because of a childhood accident she couldn't have children. In truth, she'd had her tubes tied. More than 10 years passed before she confessed her past. "His eyes filled up," she recalls. "What he was disturbed about was the fact that I had carried the burden by myself."

The marriage survived, and Blanchard went on to search for her sons, finally finding them in 1979. In the happy epilogue, the family was reunited, the secrets were told and, almost miraculously, her singing voice came back.

A single, toxic secret may skip one generation, then pop up to plague the next. Home in Georgia during a Christmas break from college in 1991, John Brown (not his real name) got an unusual Christmas card from his grandmother Elsie (not her real name), whose lunch invitation he had passed up the last time he was home. On the blank side of the card Elsie had written a note confessing that she had had "a relation" with a man named "Mike McMurphy." She had married another man, "George Brown," at a time when she thought she might still pass the child off as his. That child was John's father, still living, and without a clue as to who his own real father had been. "Needless to say, this was a shock," John says.

Over lunch, Elsie told John she felt she had been forgiven long before for what she did. "I'm, like, 'Well, if you feel forgiven, why do you need to tell it again?'" Elsie explained that if John had children, she didn't want him to worry about one of them turning out like his schizophrenic aunt—one of two children she had with the man she married—who was in and out of hospitals and often violent. "She said, 'I'm telling you for your sake, not your father's.'"

Revealing the truth brought Elsie and John closer. Two years later she asked him if he had told his father yet. He replied: "What's the point?" By then, both George Brown and the man Elsie had the affair with had been dead for years. Not that his father couldn't handle secrets. John has told him, for instance, that he is bisexual, maybe even gay. His father "reacted oddly" at first, later confessing he'd been attracted to men a few times himself.

There was a time when people were ostracized for such breaches of orthodoxy. "We've come a long way, because most people understand these problems," says Dr. Kate Wachs, a family therapist in Chicago. Yet she recognizes a continued penchant for hushing up about everything from the eccentric uncle to the horse-rustling great-grandparent. "When things are hidden," Wachs says, "they become so much more powerful. Each time you lie about it, you are telling yourself that it is a horrible thing. And when you look at it, usually it turns out that it is not so bad."

Or necessary. Stan Weinberger found out at 50 that the man he and his brother knew as their stepfather was actually his father. His mother had lied about the father's true identity because she'd had an affair with him while he was married to another woman. To this day, his mother says about keeping the secret: "We didn't do such a terrible thing." But it's a terrible thing to lie to your children, says her son, who remains saddened and a little bitter about not having known he had a father all these years. "Secrets don't work," he says. "Eventually the truth comes out, and all there is left is the pain."

Reflections

1. Is there a family secret that you have kept silent?

2. What reasons do you have to hold onto it?

3. If you chose to tell someone who might that be?

4. How might you feel as a result of exposing this secret?

11

In this reading, which first appeared on the World Wide Web, the author describes how traditional courtship is changed when it is replaced with wired correspondence. For some, electronic communication is preferable to voice because it is durable.

Wired Love: Courtship in the Age of E-Mail

By Lisa Napoli

Alissa Bushnell did what any normal, red-blooded woman would do after she had a fun date with a great guy. She freaked out, commiserated with her friends at work, then waited for him to e-mail her.

Flash Rosenberg ended a long-distance relationship with a man who used e-mail to keep in touch between visits but never communicated anything of substance. "He wouldn't talk about anything real," she complains. "It was gesture before communication."

A male friend, new to e-mail, meets a woman in a bar, and after an evening of intense conversation, she gives him her business card. Emboldened by the technology, he sends e-mail asking for a date.

"I've never asked someone out this way," he confesses, after telling her what a pleasure it was to meet her. "Did I break any rules?" No, she replies — but no, she won't go out with him.

Another single male friend says that this is exactly why he only asks women out via e-mail. "It's not as awkward as a phone call," he says. "Or as difficult if they say no."

This is not another tale of people who meet online and fall in love before ever having a dinner date, a well-documented 90s phenomenon that invariably elicits rolled eyeballs and cautious admonitions ("How can you know who you're really talking to?") from those who haven't partaken. If you dismiss this trend, you probably also don't realize that at least an eighth of the couples you know met through some form of personals advertising.

No, the subject at hand is traditional courtship in the age of e-mail, how wired correspondence changes and affects the dance of blossoming love. If you don't believe that's true, you've probably had a steady mate for the last five years or so. Or maybe you simply haven't yet realized the power of a simple electronic message to make a recipient swoon — or run in the opposite direction.

Love letters are as old as the written word, of course, and as complex and individual as lovers themselves. They are a filter, a mask, a measured way to communicate, and unless intercepted in illicit situations, they're generally safer than baring the soul face-to-face.

One of modern history's more ardent love affairs took place entirely by mail. The playwright George Bernard Shaw spoke from experience when he remarked, "The ideal love affair is one conducted by post." For years, he traded passionate correspondence, sometimes daily, with a celebrated actress of the time named Ellen Terry ("my Ellenest Ellen"). Both were married; in fact, Shaw once noted that his pen-mistress tired of "five husbands, but never of me."

When they did meet, briefly, several years into their affair of words, Terry was disappointed and the letters stopped for a while, never to regain their intensity. "He was quite unlike what I had imagined from his letters," she remarked.

"Only on paper," declared Shaw, "has humanity achieved glory, beauty, truth, knowledge, virtue and abiding love."

But then, all Shaw had was paper.

Even the most stubborn Luddite would be hard pressed to deny that there is something sensuous about a declaration of affection directed at you, irrespective of the medium that delivers it. Woe to the writer who misspells or errs grammatically; joy to the Cyranos among us.

Alissa Bushnell is a modern Roxanne who is almost as effusive about love in the age of e-mail as she is about the new man in her life. For their first month of dating, after meeting at a party, they never talked on the phone, communicating instead by e-mail. In fact, that's how he first asked her out.

"Prior to dating Cory, I was dating a man who was annoyed by my use of e-mail to communicate with him," Bushnell said. "He wanted to talk on the phone, but I would not speak with him on the phone unless he called me — The Rules, right?" she says, alluding to the best-selling book that is the Bible — and bane — of many a single woman.

"So, if I wanted to communicate something to him I would e-mail him. He found it too impersonal. I think that e-mail changes the rules of courting, particularly because it levels the gender playing field. The rules of traditional dating don't apply to the world of e-mail. We can answer and respond as thoughtfully and as strategically as a guy calling. See, when a guy calls us, like The Rules say they should, we are automatically in a reactive mode, versus a proactive stance."

In the very early stages of their relationship, Bushnell would eagerly await Cory's e-mail the way new lovers have long anticipated letters or phone calls. Once, after he had been away on a weekend camping trip, she arrived at the office on Monday morning to find that an e-mail, sent in the middle of the night, awaited her. "If a guy got back from camping at 1 A.M. and wrote you an e-mail, you know he's thinking of you," she says.

Flash Rosenberg, who lives in New York, romanticized the inconvenience of falling in love with a man in Los Angeles in the 90s. What luck, she thought, that she lived in the age of e-mail, and that 3,000 miles could seem a bit closer between visits.

But she knew something was amiss, and she felt rejected, when the man she was seeing took her off his mailing list of jokes — even though the jokes annoyed her.

"It was such a power kind of thing," she says. "I could never really talk, and he loved e-mail because it was up to him. He was so interested in new forms of communication that he couldn't communicate.

"I was female, he was e-mail."

Of course, this is not the first time a woman has complained of a man avoiding serious conversation. But to Rosenberg, this pathos took on an elevated dimension in the digital world. "When we were together, it was skin, touch, sex, promises," she says. "And then when he wrote to me, he would just tell jokes."

Even so, "once burned, twice shy" has not colored her expectations. Rosenberg says she can't imagine falling in love with an unwired man, explaining, "I wouldn't be attracted to anyone who isn't interested in how the world communicates."

There are no absolutes, of course, when it comes to matters of the heart, except this: we are guaranteed to fall in love, and out, and in again. Part of the thrill and the frustration is the unpredictable nature of circumstance and timing. And, for better and for worse, a new part of the dance is this new element of communication.

In honor of this holiday in which we celebrate love, print out your love e-mail — even your break-up e-mail — for the sake of posterity and personal archive.

Twenty years from now, couples will reminisce wistfully about such missives, much as one friend did the other evening about love letters she received from her husband when they first met 13 years ago.

"I keep them in a shoe box," she said, almost breathlessly. "I read them from time to time, and it's wonderful. It's like going back to another time, even though when I read them, I have the benefit of knowing Mark for all these years now. Somehow what he said then just seems more true."

Though much has been made of the fleeting nature of electronic communications, Bushnell sees wired love as durable.

"With words we just have the memory of what's said," she offers. "With e-mail, it's more permanent."

Reflections

1. Have you used e-mail as a means to correspond with others?
2. Have you ever had or known anyone who had a romantic relationship using the computer?
3. What advantages and disadvantages can you cite?
4. When and why would a person use this means to communicate?

12

In addition to issues of racism and assimilation, black single women have to deal with the same problems as white single women. While discussing these, the author questions the place of ethnocentric identity.

The Dilemma of Black Single Women

By Gloria Naylor

The Chinese restaurant is in a section of Pennsylvania Avenue in Washington where the table linen is changed after each meal, the Boston ferns are real, and the stir-fried beans and ginger will cost you the down payment on a plot of land to grow your own. The young woman who comes in and sits alone could have stepped out of any Ebony Christmas ad for Courvoisier: her shoulder-length braids are impeccable, the honey-brown skin has red highlights around the cheekbones from either the chill wind or conservatively applied blush. She moves her manicured hands with the grace of someone who is accustomed to having them pampered each week. As she places her leather portfolio on the seat beside her and picks up the menu, her body language is indisputable: she is treating herself in celebration of some personal victory—a successful business meeting? A new promotion at work? Bank approval of a condominium loan? Any of those could be plugged into this picture realistically, and what is all too real is the fact that she's celebrating alone.

Now, it's quite possible that young woman elected to be eating there by herself, or that she had a male counterpart waiting somewhere with yellow roses and a bottle of fairly decent wine to cap off her achievement, but the statistics and a random sampling of my friends tell me that it is not the likely case. I don't have to go to the census reports to find out that stable relationships between black men and women are quickly losing ground as the divorce rates, households headed by women and the numbers of never-married women proliferate; I can just start flipping through my telephone book. In the B's is a divorced friend who just decided to adopt a child alone, in the C's one in her early 40's who has never married, in the L's another divorced friend who went to St. Croix to find a new husband, in the M's three women raising a sum total of eight children single-handedly. Were I to ring them up, the personal histories and causes for their situations would vary, but their bewilderment would be consistent and echo my own. What exactly *is* the problem between us and our men today?

It is the same problem that is at the root of deteriorating relationships in all America, and then some. But it's the "then some" that compounds our predicament. While dealing with the pressures of modernization, along with the transitory nature of the nuclear family that all American couples must deal with, we must also battle the erosive factors that racism thrusts into the landscape of our personal relationships. One is a factor of numbers: we are a minority community overrepresented in the pools of the unemployed, the poorly educated and those subsisting below the poverty line, which will concomitantly insure overrepresentation in teenage pregnancy, drug abuse and the prison population. In short, there just aren't enough viable mates out there for black women and bleak prospects of a greater percentage being born. And this shrinking pool of prospects means that a vicious cycle is setting in for the future: our women will marry later, if at all, thus having fewer children or none. True, smaller families are occurring in the white middle-class community as well; but the loss of numbers has deeper implications for a minority community whose majority is confined to an underclass.

The second factor raises a very painful subject: one of the high costs of assimilation for the few has been the erosion of a healthy ethnocentric identity. Black professionals have found the rewards are greater for those who can swim the

---

Reprinted by permission of Sterling Lord Literistic, Inc. Copyright © 1986 by Gloria Naylor.

fastest and easiest with the flow, which means superimposing an alien set of values upon a distinct racial personality. Assertive, self-directed women were traditionally looked upon as valuable assets in African and African American families; it was a matter of common sense and survival in heavily agricultural existences and the subsequent low-paid urban ones. It gave rise to a breed of women who are conditioned to serve a functional rather than an ornamental role for their male counterparts. The concept is relatively new for us that a woman who insists on equal (or even superior) weight being given to her opinions, her salary and her decision-making ability in the home is "threatening" to black manhood.

Assimilation has not stopped with the adoption of a life-style by the leaders of the power structure; our men have adopted the upper-middle-class perceptions of an "ideal" balance between the sexes, too. Thus, professional black women find themselves having to juggle the pressure of their own careers with tense personal relationships because of expectations that they be something they simply are not—but more importantly, cannot be. It hurts to think that one of the impediments to stability with our men might be, at its most basic level, everything that is *us*.

I know that consolation cannot be found by going back to my phone book and ringing up my white female friends. They have their own horror stories to tell. They may be in a slightly larger boat, but it's still rocky.

I tend to look with a jaundiced eye upon the businesses that have grown out of this search for compatible partners: computer dating services, singles' travel clubs and newsletters. It is so typically American to find a way even to package loneliness—the construction of smaller apartments, the canning of soup for one. The "personal" columns, once confined to the back pages of smut magazines as the province of men who didn't have the courage to ask a woman to her face if she liked having hard-boiled eggs bounced on her navel, have found a new respectability and a new clientele.

Not surprisingly, ads, along with magazines like *Chocolate Singles,* are gaining popularity among upwardly mobile black women. They replace the personal networks once provided by church attendance and family and friends, now lost to individuals who are demographically and socially in transit. Such publications, composed of a few pages of newsprint with slick covers, attempt to disguise the fact that they are primarily a clearinghouse for personal ads. Tongue-in-cheek names like *Chocolate Singles* are emblematic of this subterfuge. Why not call it "Dark and Desperate," getting straight to the point? In the December issue the personals section bore out the living reality: the "Male Advertisers" took up three and a half columns, the "Female Advertisers" twice that many.

Professional black women using these ads suggest that an increasing number are finally deciding to broaden their options: the phrases "any nationality," "open to a variety of cultural experiences" and "race unimportant" appear. I happen to be in the camp that believes race and culture hold a great deal of importance in compatibility. Yet, while the natural inclination of human beings is like to like, the most unnatural thing in the world is to go through life without feeling appreciated and loved. Unlike the alarmists, I don't see this as portending the demise of our ethnic uniqueness. I have only painted one side of the scenario; the statistics also bear out that for each black woman who is alone, there is one in a stable relationship. Among my acquaintances are many families who are a living counterpart (if a much less saccharine version) of "The Cosby Show." Depending upon how you look at it, the glass is either half-empty or half-full, but either way is far from sufficient.

Reflections

1. What is the author's assessment concerning black single women?

2. How do the issues of racism, ethnocentrism, and assimilation that exist among black single women compare with your own?

3. Have you ever felt the victim of sexism, racism or any other prejudice?

4. What was the author's assessment of assimilation?

In this classic skit from the television series "Saturday Night Live," Gilda Radner, Laraine Newman, Jane Curtin, and Madeline Kahn portray twelve-year-olds discussing sex.

The Slumber Party

From "Saturday Night Live."
By Marilyn Miller

(A darkened living room with single lantern-type light used for camping. Girls huddled around Madeline on the floor with pillows, blankets, etc. Assorted old pizza boxes, coke bottles, strewn around them.)

MADELINE: (enormously confidential) ... so then, the man gets bare naked in bed with you and you both go to sleep which is why they call it sleeping together. Then you both wake up and the man says, "Why don't you slip into something more comfortable"—no, wait, maybe that comes before—it's not important—and then the man says . . . (light goes on at top of staircase)

MOTHER'S VOICE: Gilda, it's five A.M. When does the noise stop?

GILDA: We're just going to sleep, Mother.

MOTHER'S VOICE: What are you talking about at this hour?

GILDA: School!

MOTHER'S VOICE: Well, save it for the morning. (Door slams. Lights out.)

JANE: (to Madeline, as if nothing has happened) And then the man . . .

MADELINE: Anyway . . . (Brings girls closer, whispers something inaudible. We finally hear:) . . . then the man (whispers) in you and then you scream and then he screams and then it's all over. (Moment of silence. The girls sit there shocked and horrified)

LARAINE: (making throwing-up sounds, pulling blanket up over her head) That's disgusting!

GILDA: You lie, Madeline.

MADELINE: Cross my heart and hope to die. My brother told me in my driveway.

GILDA: Your brother lies, Madeline.

MADELINE: No, sir.

JANE: Come on. Isn't he the one who said if you chew your nails and swallow them a hand will grow in your stomach?

MADELINE: Well, it's also true because I read it in this book.

JANE: What'd it say?

MADELINE: It said, "The first step in human reproduction is . . . the man (whispers)

LARAINE: (hysterical, coming out from under covers) It's disgusting!

(Laraine, Gilda and Jane all do fake throwing up)

MADELINE: It's true.

JANE: Well, I just know it can't be true because nothing that sickening is true.

MADELINE: Boogers are true.
(the girls consider this for a moment)

GILDA: Well, I mainly don't believe it because I heard from my sister about this girl who this guy jumped out from the bushes and forced to have a baby.

MADELINE: (smugly) How?

GILDA: I don't know. I think he just said, "Have a baby right now."

MADELINE: Oh sure, Gilda. And you think that would work if I tried it on you?

GILDA: (scared) Hey, don't. O.K.?

© 1976 by National Broadcasting Company, Inc.

MADELINE: Well, don't worry. It wouldn't because that's not how it's done. How it's done is . . . the man . . .

LARAINE: Don't say it again, O.K.? I just ate half a pizza, O.K.?

GILDA: (thoughtfully) So that's why people were born naked.

JANE: Yeah.

LARAINE: But how could you face the man after? Wouldn't you be so embarrassed?

JANE: I'd have to kill myself after. I mean, I get embarrassed when I think how people standing next to me can see inside my ear.

MADELINE: Well, that's why you should only do it after you're married. Because then you won't be so embarrassed in front of your husband after because you're in the same family.

LARAINE: Oh, well, I really want to get married now. Not!

MADELINE: But the worst thing is—our parents do it, you know?

GILDA: Come on!

MADELINE: Gilda, think: none of us would be here unless our parents did it at least once.
(Moment of silence. They all consider the horror of this)

JANE: (horrified) My parents did it at least twice. I have a sister.

GILDA: (greater horror) And my parents did it at least three times. I have a sister and a brother.

(they all turn to give her a "you're dirty" look)

GILDA: But, like, I know they didn't do it because they wanted to. They did it because they had to. To have children.

MADELINE: (accusing) They could have adopted children.

GILDA: Yeah, but adopted children are a pain. You have to teach them how to look like you.

LARAINE: Well, my father would never do anything like that to my mother. He's too polite.

MADELINE: My father's polite and we have six kids.

LARAINE: He's obviously not as polite as you think. (they glare at each other)

JANE: I wonder whose idea this was.

MADELINE: (offhand) God's.

JANE: Oh, come on. God doesn't go around thinking up sickening things like this for people to do.

GILDA: Maybe God just wants you to do it so you'll appreciate how good the rest of your life is.

JANE: Maybe.

LARAINE: (to Madeline) How long does it take?

MADELINE: Stupid! That depends on how big the girl's stomach is and how fast she can digest.

GILDA: Oh.

JANE: Can you talk during it?

MADELINE: You have to hold your breath or else it doesn't work.
(various vomit-sounding shrieks, screams, etc.)

JANE: Well, I'm just telling my husband I'm not going to do it. (to heaven) Tough beansies.

MADELINE: What if he says he'll get divorced from you if you didn't do it?

(the girls consider this)

JANE: I would never marry someone like that.

MADELINE: What if you did by accident? What if . . . (making up story) . . . you met him in a war and married him real fast because you felt sorry for him since he'd probably get killed only he didn't and then you were stuck with him?

GILDA: (moved by emergency) Look—let's make this pact right now that after we get married, if our husbands make us do it, we'll call each other on the phone every day and talk a lot to help keep our minds off it, like our mothers do.

JANE: Right.

MADELINE: Right.

LARAINE: Right, because it's disgusting. (makes some throw-up sound. Ducks under covers)
(Laraine turns out flashlight)

JANE: Well, don't worry, we'll never have to keep this pact because I'll never do it.

GILDA: Me, neither.

MADELINE: Me, neither.
(there is a beat)

LARAINE: (quietly) I might.
(fade out)

Reflections

1. At age twelve, what were your beliefs concerning sexuality?

2. As you grew up, what were your sources of information?

3. What did your parents tell you about sex? When did they tell you?

4. Will you teach (or have you taught) your children in the same way?

5. At what age do you think children should be told about such things as reproduction, masturbation, homosexuality, the relationship between love and sex, sex as pleasurable activity, and sexually transmissible diseases? Ideally, where should they get this information?

14

Well-known author and sex therapist Marty Klein suggests that the question "Am I normal?" is one of the most pervasive in our culture.

What Is Normal?

Excerpt from *Ask Me Anything*
By Marty Klein

Why is the most common sexual question "Am I normal?"

Because unlike other important activities, we don't observe others having sex, don't hear anyone discussing their sexual feelings or experiences seriously, and don't have access to reliable information about what other people feel and do.

This fact, coupled with the pressure to be as sexually liberated and satisfied as the Joneses, leads many people to wonder if they are sexually normal.

I see all these magazine surveys, and the sexual behavior of their readers never sounds like me. Does this mean I'm abnormal?

It may mean that you express your sexuality differently than those readers do, which is not the same as being abnormal.

Consider, too, that people lie on sex surveys, generally in the direction of what they think the survey takers want to hear. People also shade their responses in the direction that they like to think of themselves. You may do this yourself when the doctor asks you your weight; we tend to answer a little closer to our ideal than we actually are.

Sex surveys are usually designed by statisticians, not sexologists. They use expressions like *make love, foreplay,* and *how many times,* which are much too simplistic to describe real people's sexuality. Though they do the best they can, respondents usually paint a simpler picture of their sexual experiences than really exists.

Finally, remember that only a small fraction of any group answers surveys, and it is not a random sample. Thus, large segments of most groups are underrepresented by surveys, mail campaigns, and the like. This is obviously what happens, for example, when a small unhappy group protests sex on television, while no one from the satisfied majority writes in praising it.

What about morality? Doesn't that figure in deciding what's normal?

Morality and normality are similar concepts. They are judgments, not facts; they cannot be disproved or refuted; anyone can decide how they should be defined for everyone; and those who don't agree are frequently criticized as inadequate human beings.

As a human activity, sexuality, by definition, has a moral component. Too frequently, however, sexual morality has stood for rigidity, limitations, and the social and political status quo. It has had less to do with honorable relationships between people and more to do with which body part goes where, and with the gender, race, age, and class of the people to whom the body parts belong. This is, at best, well-intentioned paternalism and, more commonly, bigotry. It is not "morality."

I believe the basis for a moral sexuality is simply an extension of the Golden Rule: Do not do unto others as you would not have others do unto you; and don't do it to yourself, either. The specific way you choose to implement this is up to you and is virtually irrelevant.

I rarely hear the media talk about what's normal, so why do you say they are so important in defining it for us?

The key message the media send us about sexual normality is their remarkably uniform version of sex. On television, in movies, and on magazine covers, sexy people are almost invariably young, beautiful, and eager. They have no communication problems, no concerns about birth control or disease, and they have wildly responsive bodies. No erection problems either, no sir.

The advertising industry also shapes our sexuality with a single theme, repeated millions of times during each of our lives: We are all in danger of being sexually inadequate; this is terribly painful, and there's only one dependable cure.

If we believe the advertising industry, this cure is not communication, nor resolving fears of intimacy, nor correcting early teachings or negative beliefs about sex. The solution to sexual inadequacy is buying things! Whatever the product, advertising can sexualize it—by making it a solution to a problem.

Ads tell us that what's normal is to worry about sexual adequacy and to resolve that worry by buying things. If you don't worry, you're not normal. And if you don't buy, or don't want to buy, you're also not normal.

Finally, let's remember that there are rules about what sexual images the media is willing to carry. Many women's magazines won't include information about real female sexuality—for example, the fact that women use vibrators or touch themselves during intercourse. Radio stations won't carry even tasteful ads for condoms. And television programs rarely show us gay people, except as fools or psychopaths. The trend is so uniform it can only be called censorship. Censorship of the real in favor of an invented concept of what's normal.

So how do I know if I'm sexually normal?

This, of course, is the most common sexual question. It's also the saddest, I think, because it reflects modern Americans' loss of trust in their own sexuality.

Unfortunately, most of us do not know, deep in our bones, that our sexuality is okay. Understandably, we feel continual pressure to figure out if we're sexually normal.

But healthy human sexuality has a dark side—not a bad side, just a dark one. In addition to its lighter side, our sexuality includes greedy, aggressive, lusty, controlling, amoral (not immoral) aspects. This dark side can scare people who don't trust themselves sexually, making them wonder how they can be "normal" when they have thoughts, feelings, and even behaviors that seem "bad."

How do you know if you're sexually normal? You decide. You understand that healthy sexuality is a vast, complex web of light and dark that is not dangerous. You realize that "normal" is simply a judgment and that you are as qualified to make that judgment as anyone else. So make it.

And if you commit part of your existence to expressing your sexuality in life-affirming ways, and if you really experience yourself when you do that, you don't have to decide. You feel it, and that's how you know.

Reflections

1. Have you ever wondered if you were sexually "normal"?

2. What aspects of your sexuality, if any, have caused you worry or anxiety?

3. Have you been able to resolve these issues? How?

4. Have you discussed your concerns with anyone? If not, can you think of someone—such as a friend, therapist, or group—who might be of assistance to you?

15

In this article, Newsweek *writer Sharon Begley discusses the issue of gender differences in response to infidelity. It seems that whereas men are most distressed by sexual betrayal, women are more upset by emotional infidelity. The author offers some theories as to why this may be so.*

Infidelity and the Science of Cheating

By Sharon Begley

Think of a committed romantic relationship that you have now, or that you had in the past. Now imagine that your spouse, or significant other, becomes interested in someone else. What would distress you more:
- Discovering that he or she has formed a deep emotional attachment to the other, confiding in that person and seeking comfort there rather than from you?
- Discovering that your partner is enjoying daily passionate sex with the other person, trying positions rarely seen outside the Kamasutra?

While this makes for an interesting party game—though we don't advise trying it around the family Christmas table—the question has a more serious purpose. Researchers have been using such "forced choice" experiments to probe one of the more controversial questions in psychology: why do more men than women say sexual betrayal is more upsetting, while more women than men find emotional infidelity more disturbing? Psychologist David Buss of the University of Texas, Austin, first reported this gender gap in 1992. Since then other researchers have repeatedly found the same pattern. But when it comes to explaining *why* men and women differ, the battle rages.

The year now ending brought claims that genes inherited from our parents make us risk takers or neurotic, happy or sad. In the new year, watch out for ever more studies on how genes passed down from Neanderthal days make us what we are. "There is tremendous interest in evolutionary perspectives in psychology," says John Kihlstrom of Yale University, editor of the journal Psychological Science. And not just among scientists. In 1996, magazine articles waxed scholarly on how evolution explains, for instance, Dick Morris's extramarital escapades. Basically, his DNA made him do it.

The debate shapes up like this. Evolutionary psychologists argue that sex differences in jealousy are a legacy of humankind's past, a biological imperative that no amount of reason, no veneer of civilization can entirely quash. In other words, genes for traits that characterized the earliest humans shape how we think, feel and act, even if we are doing that thinking, feeling and acting in cities rather than in caves. In particular, men fly into a rage over adultery because to do so is hard-wired into their genes (not to mention their jeans). The reason is that a man can never be altogether sure of paternity. If, at the dawn of humanity, a man's partner slept around, he could have wound up inadvertently supporting the child of a rival; he would also have had fewer chances of impregnating her himself. That would have given him a poor chance of transmitting his genes to the next generation. Or, put another way, only men who carried the gene that made them livid over a spouse's roaming managed to leave descendants. Says UT's Buss, "Any man who didn't [do all he could to keep his wife from straying sexually] is not our ancestor."

For a woman, the stakes were different. If her partner sired another's child, his infidelity could have been over in minutes. (OK, seconds.) But if he became emotionally involved with another woman, he might have abandoned wife No. 1. That would have made it harder for her to raise children. So women are evolutionarily programmed to become more distressed at emotional infidelity than sexual infidelity.

From Newsweek, 12/30/96. © 1996, Newsweek, Inc. All rights reserved. Reprinted by permission.

The journal Psychological Science recently devoted a special section to the controversy. Leading off: a study by Buss, working with colleagues from Germany and the Netherlands, in which 200 German and 207 Dutch adults were asked the standard "which is more upsetting" question. As usual, more men than women in both cultures said that sexual infidelity bothered them more than emotional infidelity. "This sex difference is quite solid," says Buss. "It's been replicated by our critics and in cross-cultural studies, giving exactly the results that the evolutionary theory predicts."

Critics of the evolutionary paradigm say it is dangerous to call the jealousy gender gap a product of our genes. "This theory holds profound implications for legal and social policy," says psychologist David DeSteno of Ohio State University. "Men could get away with murder [of a sexually unfaithful spouse] by attributing it to their biology and saying they had no control over themselves." What's more, he argues, the theory is wrong. First, if there are genes for jealousy, they can apparently be influenced by culture. Although in every country more men than women were indeed more upset by sexual infidelity than the emotional variety, the differences between the sexes varied widely. Three times as many American men than women said that sexual treachery upset them more; only 50 percent more German men than women said that. The Dutch fell in between. So the society in which one lives can change beliefs, and thus make the gender gap larger or smaller.

More problematic for evolutionary psychology is another repeated finding. Yes, more men than women find sexual infidelity more disturbing. Something like 45 percent of men and 10 percent of women, or 30 percent of men and 8 percent of women (the numbers depend, says Buss, on how the question is worded), were more upset by the idea of sexual betrayal. But look more closely at the numbers for men. If 45, or 30, percent say that sexual betrayal disturbs them more, that means that most (55 percent, 70 percent) are *not* disturbed more by it. Yet evolutionary theory predicts that, even though men should not be indifferent to emotional infidelity, they should care more about the sexual kind.

Scientists who have been skeptical about the "my genes made me think it" theory have a different explanation for the jealousy gender gap. What triggers jealousy depends not on ancient genes, they argue, but on how you think the opposite gender connects love to sex and sex to love. Or, as psychologists Christine Harris and Nicholas Christenfeld of the University of California, San Diego, propose, "reasonable differences between the sexes in how they interpret evidence of infidelity" explain the gender gap. In other words, a man thinks that women have sex only when they are in love; if he learns that a woman has had sex with another man, he assumes that she loves him, too. Thus sexual infidelity means emotional infidelity as well. But men believe also that women can be emotionally intimate with another man without leaping into bed with him. A woman's emotional infidelity, then, implies nothing beyond that. By this reasoning, men see sexual betrayal as what Peter Salovey of Yale University and OSU's DeSteno call a "double shot" of infidelity. Sexual infidelity is therefore more threatening than mere emotional infidelity.

A woman, on the other hand, notices that men can have sex without love. Thus a man's sexual betrayal does not necessarily mean that he has fallen in love with someone else. So adultery bothers her less than it does men. But a woman also notices that men do not form emotional attachments easily. When they do, it's a real threat to the relationship. Says DeSteno, "Whichever type of infidelity represents a double shot would bother someone more."

Now scientists are designing experiments to show whether the mind's ability to reason, rather than genes, can explain the jealousy gender gap. The UCSD team asked 137 undergraduates the "which distresses you more" question. As expected more men than women picked sexual infidelity as more upsetting. But the researchers also found differences in men's and women's beliefs. Women thought that, for men, love implies sex more often than sex implies love. And men said that, for women, sex implies love about as strongly as love implies sex. This difference in assessments of the opposite sex, argue the UCSD psychologists, explains all the gender gap in jealousy. Of course a woman is more bothered by a man's emotional infidelity than by sexual betrayal: a man in love is a man having sex, they figure, but a man having sex is not necessarily a man in love. Now there's a shock.

Other experiments undermine as well the "my genes made me think it" argument. DeSteno and Salovey asked 114 undergraduates, and then 141 adults ages 17 to 70, how likely it is that someone of the opposite sex who is in love will soon be having sex, and how likely that someone of the opposite sex who is having sex is or will be in love. Anyone, man or woman, who believed that love is more likely to mean sex than sex is to mean love was more upset by emotional infidelity than by sexual infidelity. And anyone, man or woman, who believed that someone having sex is someone in love found sexual infidelity more upsetting. These data, says DeSteno, "argue against the evolutionary interpretation. Which infidelity upsets you more seems related to [gender] only because [gender] is correlated with beliefs about whether sex implies love and love implies sex."

Evolutionary psychologists don't buy it. Buss points to studies showing that a woman is at greatest risk of being battered, and even murdered, by her partner when he suspects her of sexual infidelity. "Men's sexual jealousy is an extremely powerful emotion. It makes them go berserk," says Buss. "The 'rational' arguments don't square with [the fact that] jealousy feels 'beyond rationality.' This vague implication that culture and socialization [cause sex differences] is very old-social-science stuff that sophisticated people don't argue anymore ... Sometimes I feel that I am amidst members of the Flat Earth Society."

For all the brickbats being hurled, there is some common ground between the opposing camps. Buss and colleagues believe that jealousy, like other emotions sculpted by evolution, is "sensitive to sociocultural conditions." And those who scoff at evolutionary psychology agree that, as DeSteno says, "of course evolution plays a role in human behavior." The real fight centers on whether that role is paramount and direct, or whether biology is so dwarfed by culture and human reason that it adds little to our understanding of behavior. Spinning stories of how Neanderthal genes make us think and act the way we do undeniably makes for a lively parlor game. (Example: men prefer women in short skirts because they learned, millennia ago on the savanna, that women in long skirts tended to trip a lot and squash their babies.) And it is one that will be played often in 1997. If there is a lesson here, it may be this: be wary of single-bullet theories advanced so brilliantly that their dazzle gets in the way of their content.

Reflections

1. Have you experienced sexual or emotional infidelity by a partner?

2. Have you experienced jealousy in a love relationship? If so, what caused it? Was your jealousy "suspicious" or "reactive"?

3. In your own experience, do men and women react differently depending on the type of betrayal? How?

4. Do you think your own responses are influenced more by biology or by culture? Can you give examples?

5. Did reading this article change your way of thinking about infidelity? In what way?

16

Marcie Rendon, a Native American midwife, recalls the births of each of her three daughters.

A Native American Birth Story

From *Birth Stories* (edited by Janet Isaacs Ashford)
By Marcie Rendon

A Native American birth story. Is it any different than any woman's birth story? Any mother, the world over, could come upon a woman in labor anyplace and recognize what was happening. And feel at one with, be supportive of, the woman in labor.

Ever since I can remember, I knew I wanted to have my babies at home. There was never a question in my mind whether I would have children or not. I just knew I would and that I wanted them born at home. I figured, women have given birth since time began, our bodies know what to do and how to do it.

There isn't a birth story. There are birth stories. Mine began with my asking, "Grandpa, tell me again, how were you born?"

"In the snow, girl, dropped in the snow. My ma didn't know nothin. Raised in mission schools. Didn't know nothin. Just dropped in the snow. Till my grandma heard her cryin and came out and found us. Thought she had to go to the bathroom; went out and dropped me right in the snow. Raised in mission schools. Didn't know nothin.

"Now your ma, she was born the old way. In the fall, during wild rice season. Right there in ricing camp. Your grandma and me were ricing when your ma started to be born. Three old ladies helped her. They stuck sticks in the ground like crutches. Your grandma squatted, hanging on them under her arms. They had medicines for her. Right there. Knew what to do to help your ma be born. Yep, right there in ricing camp, your ma was born. The old way, no trouble. Me, I was dropped in the snow. Thought she had to go to the bathroom. Didn't know nothing. Raised in mission schools."

My birth. Born to an alcoholic mother. Suffering withdrawal along with her during the hospital stay. Nursing, etched in my being as a wordless memory. Remembering her story, laughing. "When I was nursing I had so much milk. My breasts were so big I'd take bets in the bar. I'd take bets that I could set a glass of beer on my tits and drink the foam off. Got a lot of free drinks that way." Laughter. Laughter that filled the whole room. Caught everyone up in it. My mother, dead in my eighteenth year, from alcohol.

Birth stories. My daughters, children of my grandfathers' and grandmothers' dreams. You were a reality long before my passion and your father's seeds conceived you on those winter and spring nights. Visions preceded flesh from my womb. And my generations shall be as one. Spirits healing my mother's madness, my father's sadness. As the daughters of my grandfathers' and grandmothers' dreams touch hands with wind and water, light and love.

"Tell me Mom, how was I born?" asks my seven-year-old.

Conception. Christmas Eve. Warm glow as the spark of your being ignites. I remember standing at the window that night watching the lights glitter on the snow, knowing that night that you were started in me.

"Your daddy told me I was crazy because I wanted to have you at home. I asked all over, trying to find a midwife. Everyone told me there weren't any, that midwives were a hundred years ago. So I decided I'd have you at home alone. I kept saying, 'Women have given birth since time began; my body knows what to do.' And your daddy just kept telling me I was crazy. As it was, I didn't get any support to have you at home and we ended up in the hospital. They strapped me down and I had to physically fight the nurses, doctor and anesthesiologist not to be gassed or

© 1984 by Janet Isaacs Ashford. Published by The Crossing Press, 97 Hangar Way, Watsonville, CA.

drugged. Your daddy just kept telling me to be good and cooperate. I had back labor and I just kept seeing the fear in his eyes and wishing it would go away. I just wanted to have you my way, with no interference. And you ended up being born by forceps because I wouldn't cooperate. All my energy went into fighting the nurses and doctors until my body just gave up and they pulled you out. But you were a beautiful little girl. When I first got to hold you and nurse you, you just smiled right at me. They didn't want me to nurse you. They kept giving me sugar water to give to you. They told me you'd dehydrate and have brain damage if I didn't give it to you. So I'd drink it myself or pour it down the toilet to keep them off my back. When we went home I was black and blue from where the straps had been, from fighting them so hard to have you the way I'd wanted to have you." My first daughter, Rachel Rainbeaux. My "One of Many Dreams." Rainbeaux. The one who fights for life with every ounce of muscle and energy in her little body.

"And me, Mommy, how was I born?" asks my five-year-old.

"You, my girl, were born in a midwife unit. Again, I wanted to have you at home, but couldn't find a midwife to come. So the midwife unit was a compromise to your dad, who was still too scared to do it alone at home.

"You were conceived in Denver. I remember standing outside after work, waiting for your daddy to pick me up. It was April and it started to rain snow. Huge, soaking wet snowflakes. At the same time I could feel my body ovulating. That night we made love and right afterwards my whole body started to shake. I told your daddy, 'I'm pregnant. You have to get me something to eat.' He thought I was being really silly. I said, 'No, for real, get me something to eat.' So he brought me a big bowl of cornflakes. That's the night you were started.

"You were born in a hospital midwife unit. The whole birth was beautiful. Calm. Peaceful. Except I had back labor again, even though I'd made one of the nurse-midwives promise I wouldn't have back labor twice in a row. Only during transition did I lose it; the back labor got so bad I said, 'That's it. I quit, give me a cigarette, I'm done.' Everyone just laughed at me and ten minutes later you were born. Another beautiful girl who smiled right at me. I got to hold you right away. Hold you. Love you. Nurse you."

"And my foot, Mom. Tell me about my foot."

"You were born with a crooked foot. So they put a tiny little cast on it the day after you were born and then you had surgery on it when you were four months old."

Simone. My Starfire, with the warmth of a soft summer night, the flicker of eternity in her eyes. Rainy Day Woman—life giver—love flows out of your being and waters the souls of those you touch.

The next four years. Mothering. A divorce. Single parenting. Welfare. Apprenticing. Midwifing. Mothering.

"And the baby, Mom," the seven- and five-year-old say. "Where was she born? You finally got to have one of us at home, didn't you?"

My baby, four months old and growing. My home birth. What will I tell her when she's old enough to ask?

"You, my baby girl, were born at home. On the living room floor. Twenty-four hours of the hardest work I ever hope to do. Twenty-four hours of excruciating back labor. You were conceived November 2, 1982. It was right after the full moon, but it was a really cold, dark, hard night. Your dad and I had been seeing each other off and on. That just happened to be one of the 'on' nights. But the thing I remember most was that for about a month before that, every night when I'd got to bed, I'd hear three knocks on the window. I'd say, 'Bendigan, come in' like I was taught to say when the spirits are heard. Well, November 2 was the last night I heard the three knocks. Again I knew I was pregnant right away. It was so free this time, because I knew I could do whatever I wanted with the pregnancy and birth. I could have women around me to be supportive of me. I spotted a little blood in the third month, so in the seventh month we went up north to be doctored by a medicine man for a low placenta. The very first time he saw me, he said, 'Another girl, huh? Well, we'll keep prayin' for a boy.' He also told me my delivery would be dangerous, to not have anyone around who would be afraid. Well babe, you were born in the heat of the hottest summer I remember. We spent all our money all summer going where it was cool—swimming, air-conditioned shopping malls, swimming, to friends' houses that had air-conditioning, and swimming again. Your birth. Like I said, long and hard. Twenty-four hours of excruciating back labor. I walked and squatted the whole time. Trying to dance you down into a more straight up and down position. We had a grandmother, a midwife and an apprentice here to help us. Three of the most beautiful women I've ever seen. At one point I got a contraction band around my uterus, looked like a rubber band being tightened around my belly. I got scared for you. I said, 'I want to go in.' I started praying the Serenity Prayer—God grant me the serenity to accept the things I cannot change (the band, the back labor, the pain), and the courage to change the things I can (my position). I walked from the dining room to the living room and squatted down by a chair. The band disappeared and I started pushing. Hard work. Three hours of pushing during which time I visualized the Viet Nam war. Telling myself, if my brother could live through that, I could live through pushing one baby out. Just before you were born I told them I was bleeding. They said no, there wasn't any blood. Your head was born, your shoulders stuck. As soon as you were completely born the blood gushed out. Your cord was too short to nurse you or even pick you up enough to tell if you were a girl or boy. But you were lying on my belly. The first thing you did was lift your head up, look at me and smile. Another beautiful girl. They had to call another midwife to help stop the bleeding and fix the tear. And you baby, you never missed a beat. Like a turtle, your little heart never even flickered a sign of distress. Born at home. I held you, nursed you, loved you. Another beautiful, beautiful girl."

Awanewquay. Quiet, calm, peaceful little woman. Awan. Fog Woman. Three knocks announcing your intention. Birth and death inseparable. Birth—the coming of the spirit to this world. Death—the going of the spirit to that world. Fog Woman. The cloud of mystery. With your sisters I felt strong, powerful. After they were born I felt, if I can do that, I can move mountains. Well, honey, with you I moved that mountain. With all humbleness, I moved the mountain.

A Native American birth story. It is the story of the generations. I gave birth because I was born a woman. The seeds of the future generations were carried in my womb. I remember conception because the female side of life is always fertile first. I gave birth three times as naturally as possible, given the situation, because as a woman my body and heart knew what to do. I nursed because my breasts filled with milk. I remember their names because that is how they will be recognized by their grandfathers and grandmothers who have gone on before. I am a mother because I was given three daughters to love. I am a midwife because women will continue to give birth. That is my story. Megwitch. I am Marcie Rendon, Awanewquay, of the Eagle Clan, Ojibwe.

Reflections

1. Ms. Rendon asks if a Native American birth story is "any different than any woman's birth story." What do you think?

2. How does each of her birth situations reflect the prevailing attitudes of its time and place?

3. What is the story of your birth?

4. If you have a child, what is the story of his or her birth?

In this essay, the author writes of her decision to end a much wanted pregnancy by abortion. As you read, try to imagine what you would feel and do if you were in her or her husband's situation.

There is No Good Decision

By Nancy R. Nerenberg

Often, particularly among women, the subject of children and pregnancies comes up. Depending on how well I know the people, or on how I'm feeling at the moment, I will share that I had a pregnancy between Maya and Miles. Sometimes I will say that I "lost" the pregnancy. Sometimes I'll imply that I miscarried. And occasionally I will use the term that, though closest in truth, sounds so cold and hard to me: My husband and I decided to "terminate" the pregnancy.

I had no cause for concern during my second pregnancy. I had delivered a delightful, healthy baby girl a year before. Why not have another? In fact, I felt a confidence and peace of mind totally lacking during the first pregnancy. Instead of being completely neurotic about everything possible, this time I was fairly relaxed. I was a veteran. Even my second amniocentesis seemed like a familiar part of the normal routine. I proudly showed the black-and-white Polaroid of my ultrasound baby to my friends; I joked that she had my nose. And in the three-week period after the amniocentesis and before its results, I had begun to feel the "quickening," those wonderful first movements inside that tell us that new life will soon arrive.

So when I received a message to call back the doctor concerning some "unusual" results, my heart dropped through my stomach down to my feet. I couldn't believe what I was hearing.

"Something abnormal in the results...." My ears started to burn. "...In the 23rd pair, the XX or XY formation determines the sex...." I was trying to concentrate, but the doctor's voice kept moving further away. "...In your case, there is a female fetus. However, in a certain percentage of your sample cells..." I had to sit down. Even my vision seemed distorted. Objects in the room were strangely muted and at a distance from me. "...the XX cells have lost an X chromosome. It's simply fallen away, and what results is a female..." The voice continued, somehow assuming a life of its own, connected neither to me nor to the doctor speaking at the other end:

"...child with Turner's syndrome.... In mild cases what we see is a female child of small stature, usually no taller than 4 feet 9 inches or so...may need hormones to induce puberty...most likely will be sterile...in severe cases...could additionally include mental retardation as well as severe heart and kidney deformities...and perhaps a 'webbed' neck appearance, where the neck fans down and outward from the base of the head to the outside of the shoulders... In your case, however, we don't know how dramatically the child will be affected, as only a small percent of the cells, perhaps from 17 to 50 percent, appear abnormal...child could be completely normal, with no signs of Turner's whatsoever...or it could have very serious problems..."

He ended the call, saying he would like me to return in two weeks, so that he and two other specialists could view the more developed 21-week-old fetus for abnormalities and so we could sit down with a genetic counselor and decide what to do.

No clear answers

The next few weeks passed like a long, bad dream. I kept hoping I would awake and someone would tell me everything was fine, but every morning I woke to the same sickening ache in the pit of my stomach. Everyone had an opinion: Think of how it will affect a sibling; think of the costs; think of your heartache; start with a full deck. Or: I'm sure it's a normal baby; you can

Reprinted with permission of Nancy Nerenberg. © 1997

accept a less-than-perfect child; this baby is also a gift. I listened, but I continued having these feelings of not quite being present. I felt numb and detached. Reality seemed near; horribly near, and yet always a few, elusive steps away from me.

If only there were clear answers. If only someone could tell me how this child would live. I would see handicapped children on the street and flush with pain inside. I know many parents love these children as much as any other, but to me it was a tragedy just the same. Of course, if something terrible happens and you have no choice, I assume most of us will rise to the occasion. But was I willing to choose this route? Why did I decide to have an amniocentesis in the first place if not to eliminate some serious problems?

What was worst was that no one was saying, for example, "This child definitely has extremely limited potential for any semblance of a normal life." Perhaps I would have had an easier time making a decision. However, what I was being told was, in effect, "This child may be perfectly normal or severely handicapped. You decide what to do."

Two hellish weeks passed, and my husband and I went for our ultrasound/counseling appointment. They gave us no additional information, except concerning our options, including "termination." (I heard the term for the first time.) It was too late for a typical abortion procedure: I would need to be admitted, put under with general anesthesia, and have the fetus removed surgically. For this delicate procedure a specialist at Stanford was recommended. We had only a day to decide; by the next week, that doctor's next surgery day, we would be beyond his cutoff date for elective termination. "Please understand," said the geneticist kindly, "We will not judge you, however you decide. This is a terrible situation. There is *no* good decision."

My husband and I left the office and walked outside. It was a gorgeous, almost balmy fall day. The sun warmed the skin through my shirt as we walked slowly around the quiet residential block. Just a hint of warm breeze tickled the hairs on my arms. Trees were turning color and yellow leaves crackled under our feet. I wondered how nature could be so incredibly beautiful while I felt so bleak inside. We talked quietly, our thoughts going in circles. We kept looking to each other for an answer, even for any little push or shove to help us lean one direction or another. Neither of us could make that move. I've never believed that abortion was immoral; I think it's every woman's right. But this decision, which to me seemed truly a decision of life or death, was the most difficult and horrible decision I have ever had to make in my life.

At home that evening I pulled out the Polaroid of my daughter that I had been given at the first ultrasound. It is the only physical evidence of her I will ever have. I could see the profile of the face. I imagined her having my nose. My husband looked over my shoulder and said softly, "If you were standing, blindfolded, on top of a high dive, and someone told you that the pool might have water in it, and might not, would you jump?"

A special kind of grief

Suddenly something finally broke inside of me. The numbness and detachment disappeared, and a searing pain took their place. I realized I would never know this child. I threw myself face down on the bed and cried with an abandon I have never known before. I cried for me and for my baby, for all lost pregnancies and for all the handicapped children that ever were. But most of all I cried for my loss, my baby my baby my baby that was being taken away from me. Then I felt my husband's weight on the bed beside me and heard this strange choking sound, that rusty, painful wracking of someone unaccustomed to letting go. And soon this man whom I had never known to cry was heaving and sobbing with such a painful, gut-wrenching sadness. Long into the night, slowed only by pure physical exhaustion, we grieved for this baby-never-to-be.

A special kind of grief accompanies the loss of a child, even one still unborn. Nothing can touch it. The experience took my husband and me to a place where we have never been closer, but to that same place we hope never to go again. For me, as for many women, the pain would not even begin to ease until I conceived again. Becoming pregnant became an overriding drive. For my husband, the opposite was true: His pain made him avoid another pregnancy. This conflict drove us apart and threatened our relationship.

It's been six years now, and things have improved. Less than two years after our experience, we were blessed with a third pregnancy and a healthy, charming baby boy. Most of the time I am very happy. It seems trite to say, but it is true that my two children bring me more joy than I ever imagined possible. My experience, however, has changed me: A certain innocence is gone forever. People say you can't look back; sometimes I pull out my Polaroid and wonder.

Reflections

1. Have you or someone you know had to make a decision regarding abortion? If so, how was the decision reached? What feelings were experienced along the way?

2. For you, is abortion a medical issue, a moral issue, or a religious issue (or a combination of these)?

3. Do you feel it is up to the mother, the father, or society (or a combination of these) to make decisions about abortion?

4. If you found yourself in a situation similar to Ms. Nerenberg's, what do you think you would do?

18

Barbara Kantrowitz writes of the issues faced by same-sex couples who are raising children and by the children themselves. As society as a whole becomes more accustomed to the idea of gay and lesbian relationships, perhaps some of the problems faced by such families will be mitigated.

Gay Families Come Out

By Barbara Kantrowitz

There were moments in Chaire's childhood that seemed to call for a little ... ingenuity. Like when friends came over. How could she explain the presence of Dorothy, the woman who moved into her Chicago home after Claire's dad left? Sometimes Claire said Dorothy was the housekeeper; other times she was an "aunt." In the living room, Claire would cover up the titles of books like "Lesbian Love Stories." More than a decade later, Claire's mother, Lee, recalls silently watching her daughter at the bookcase. It was, she says, "extremely painful to me." Even today, Lee and Claire—now 24 and recently married—want to be identified only by their middle names because they're worried about what their co-workers might think.

Hundreds of miles away, a 5-year-old girl named Lily lives in a toy-filled house with her mommies—Abby Rubenfeld, 43, a Nashville lawyer, and Debra Alberts, 38, a drug- and alcohol-abuse counselor who quit working to stay home. Rubenfeld and Alberts don't feel they should have to hide their relationship. It is, after all, the '90s, when companies like IBM offer gay partners the same benefits as husbands and wives, and celebrity couples like Melissa Etheridge and Julie Cypher proudly announce their expectant motherhood (interview).

Lily was conceived in a very '90s way; her father, Jim Hough, is a gay lawyer in New York who once worked as Rubenfeld's assistant and had always wanted to have kids. He flew to Nashville and the trio discussed his general health, his HIV status (negative) and logistics. They decided Rubenfeld would bear the child because Alberts is diabetic and pregnancy could be dangerous. They all signed a contract specifying that Hough has no financial or legal obligation. Then Rubenfeld figured out when she would be ovulating, and Hough flew down to donate his sperm so Alberts could artificially inseminate her at home. Nine months later, Lily was born.

Two daughters, two very different families. One haunted by secrecy, the other determined to be open. In the last few years, families headed by gay parents have stepped out of the shadows and moved toward the mainstream. Researchers believe the number of gay families is steadily increasing, although no one knows exactly how many there are. Estimates range from 6 million to 14 million children with at least one gay parent. Adoption agencies report more and more inquiries from prospective parents—especially men—who identify themselves as gay, and sperm banks say they're in the midst of what some call a "gayby boom" propelled by lesbians.

But being open does not always mean being accepted. Many Americans are still very uncomfortable with the idea of gay parents—either because of religious objections, genuine concern for the welfare of the children or bias against homosexuals in general. In a recent NEWSWEEK survey, almost half of those polled felt gays should not be allowed to adopt, although 57 percent thought gays could be just as good at parenting as straight people. Despite the tolerance of big companies like IBM, most gay partners do not receive spousal health benefits. Congress recently passed—and President Clinton signed—a bill allowing states to ban same-sex marriages. Only 13 states specifically permit single lesbians or gay men to adopt, according to the Lambda Legal Defense and Education Fund, a gay-rights advocacy group. Even then, usually only one partner is the parent of record—leaving the other in legal limbo. Courts have allowed adoptions by a second parent (either gay or straight) in some of those states, although the law is still in flux. In California, for example, Gov. Pete Wilson has been lobbying hard against his state's fairly open procedure for second-parent adoptions.

Dealing with other people's prejudices continues to be a rite of passage for children in gay families. Merle, 14, lives north of Boston with her mother, Molly, and her mother's partner, Laura. Over the years she has learned to ignore the name-calling—gay, queer, faggot—from kids who know her mother is a lesbian and assume she must be one, too (as far as she knows, she isn't). And there are other painful memories, like the time in fifth grade when a friend suddenly "changed her mind" about sleeping over. Merle later learned that the girl's parents had found out about Molly and Laura and wouldn't let their daughter associate with Merle. One day in sixth-grade health class, the teacher asked for examples of different kinds of families. When Merle raised her hand and said, "lesbian," the teacher responded: "This is such a nice town. There wouldn't be any lesbians living here."

Gays say they hope that being honest with the outside world will ultimately increase tolerance, just as parenthood makes them feel more connected to their communities. "It sort of gets you into the Mom and Dad clubs of America," says Jenifer Firestone, a lesbian mother and gay-family educator is Boston. Having a child can also repair strained family relations; mothers and fathers who may have once turned their backs on gay sons and daughters often find it emotionally impossible to ignore their grandchildren.

Still, the outlook for children in this new generation of gay families is unclear. Only a few have even reached school age, so there are no long-term studies available of what the effects of growing up in such a family might be. Researchers do have some data on kids who grew up about the same time that Claire was living with Lee and Dorothy in Chicago. Most were born to a married mother and father who later split up. If the children were young, they generally wound up living with their mother, as did the majority of children of divorce. Pressures were often intense. The children worried about losing friends, while the mothers worried about losing custody if anyone found out about their sexual orientation. Yet despite these problems, the families were usually emotionally cohesive. In a comprehensive 1992 summary of studies of gay parenting, psychologist Charlotte Patterson of the University of Virginia concluded that the children are just as well adjusted (i.e., they do not have any more psychological problems and do just as well in school) as the offspring of heterosexual parents. The studies also show that as adults, they are no more likely to be gay than are children of straight parents.

The new generation of gay parents is far more diverse and will be harder to analyze. Often they are already in stable partnerships when they decide to start a family. They include lesbian couples who give birth through artificial insemination (the donors can be friends or anonymous contributors to a sperm bank); gay dads who adopt, hire surrogate mothers or pair up with lesbian friends to co-parent, and the more traditional—in this context, at least—parents who started out in heterosexual unions.

Usually they try to settle in a relatively liberal community within a large urban area like Boston, Chicago or Los Angeles, where their children will be able to mix with all kinds of families. They often join one of the many support groups that have been springing up around the country, like Gay and Lesbian Parents Coalition International or COLAGE, an acronym for Children of Lesbians and Gays Everywhere. The support groups form a kind of extended family, a shelter against the often hostile outside world.

A decade ago, when gay parents routinely hid their sexual orientation, the issues of differences rarely came up in school. But now gay parents say they try to be straightforward from the first day of class. Marilyn Morales, 34, and her partner, Angela Diaz, 37, live on Chicago's Northwest Side with their son, Christopher, 6, and their 4-month-old daughter, Alejandra, both conceived through artificial insemination. Registering Christopher for school proved to be an education for everyone. Because Morales appeared to be a single mother, a school official asked whether the family was receiving welfare. When Morales explained the situation, the woman was clearly embarrassed. "People don't know how to react," says Diaz. At Christopher's first soccer game, Diaz had to fill out a form that asked for "father's name." She scratched out "father's name" and wrote "Marilyn Morales." Both Morales and Diaz feel Christopher is more accepted now. "At birthday parties people say, 'Here comes Christopher's moms'," says Morales. Dazelle Steele's son Kyle is a friend of Christopher's, and the two boys often sleep over at each other's home. "They're such great parents," Steele says of Diaz and

Morales. "Their actions spoke louder to me than rhetoric about their political decisions."

To the parents, each new encounter can feel like coming out all over again. Brian and Bernie are a Boston-area couple who don't want their last names used because they are in the process of finalizing the adoptions of two boys, ages 12 and 6. A few years ago, Brian dreaded meeting the older boy's Cub Scout leader because the man had actively tried to block a sex-education curriculum in the schools. But his son Ryan wanted badly to join the Scouts, and Brian felt he needed to tell the man that the boy's parents were gay. As it turned out, the session went better than Brian had expected. "People challenge themselves, and people grow," Brian says. But, he adds, "as out as I am, I still feel the blood pressure go up, I sweat profusely, I'm red in the face as I tell him I'm gay, that I have a partner and that Ryan has two dads. I always think how it looks to Ryan. I'm always hoping he doesn't see me sweat."

Even in the relatively more tolerant '90s, gay parents "always feel threatened," says April Martin, a New York family therapist who is also a lesbian mother and the author of "The Lesbian and Gay Parenting Handbook." "How can you feel secure when it's still legal for someone to tear apart your family?" The parents are haunted by such well-publicized legal cases as the 1995 Virginia Supreme Court ruling that Sharon Bottoms was an unfit parent because she is a lesbian; she had to surrender custody of her 5-year-old son, Tyler, to her mother. In Florida this summer, the state appeals court ruled that John Ward, who was convicted of murdering his first wife in 1974, was a more fit parent than his ex-wife Mary, a lesbian.

Catherine Harris, 41, a university administrator in Boston, knows only too well the pain of these legal battles. Ten years ago, she was married and the mother of a toddler daughter, Tayler. Then she fell in love with Paula Vincent, now 38, a nurse-midwife. During the divorce Harris's husband fought for custody of Tayler, and Harris's parents, who disapproved of her new identity as a lesbian, testified against her. Her ex-husband won.

Harris is still on rocky terms with her parents and her ex-husband, but she and Vincent have started a new family of their own that now includes Sora, 7, and her twin siblings, Kaelyn and Marilla, 22 months. In contrast to Tayler, Sora knows her biological father only as "the donor." She has seen the vial his sperm came in and knows that her biological mother, Vincent, and Harris chose him because—according to the questionnaire he filled out at the sperm bank—he was well educated, spiritual and optimistic. " I don't really want a dad," says Sora. " I like having two moms."

But problems can arise even in the most innocent situations. Wayne Steinman and Sal Iacullo didn't truly understand their fragile footing until Labor Day weekend a few years ago, when they drove to Disney World from their home in New York City. As they passed through Virginia, Steinman was at the steering wheel; Iacullo was in the back seat with their adopted daughter, Hope, now 9. They noticed a pickup truck sticking close to them, and when they pulled off the highway to get lunch the truck followed. Just as they were getting ready to pay the bill, two highway patrolmen walked in and started questioning them. The driver of the pickup had called the cops because he suspected the fathers of kidnapping. Fortunately, Steinman and Iacullo were able to convince the patrolmen that they were, in fact, Hope's parents. "From that point on, we carried the adoption papers in our pockets," says Iacullo.

Legalities aside, gay parents—and those who disapprove of gay families—are also concerned about issues of the children's emotional development. Most same-sex parents say they make a special effort to ensure that their kids learn to relate to adults of the opposite sex. Their situation is not that different from that of heterosexual single parents, and the solution is often the same: persuading aunts, uncles or grandparents to be part of their children's lives. Hope Steinman-Iacullo, for example, often visits with her grandmother, her aunts and her teenage cousins. "There are a lot of female role models," says Iacullo.

Psychologists say the best time to tell kids how their families are different is either in childhood or in late adolescence. Young adolescents—from about ages 11 to 15—are particularly vulnerable because they are struggling with their own issues of sexual identity. George Kuhlman and his ex-wife shared joint custody of their daughter, Annie, who was 13 when their marriage fell apart in the early 1980s. But although Annie talked to her father nearly every day of her life, he never told her he was gay. "Several of my friends and even family members had been of the opinion that there might be some real psychological damage and some anger if I didn't make the disclosure," says Kuhlman, now 49 and the ethics counsel for the American Bar Association in Chicago. "That was the bear breathing down my neck." But the timing never seemed right.

Then, one day when Annie was a college freshman, he called to say goodbye as he was about to head off for a Caribbean vacation with a male friend. "She just said, 'Dad, I know. I've known for a long time ... I just thought you and Tom would have a much nicer time and a happier vacation if you know that I knew and I love you.' I pretty much fell to pieces." Annie, now 24, says she is happy she learned about her father when she was an adult. His sexuality isn't an issue now, she says. "When you have a dedicated parent, it matters less."

And, ultimately, it is the quality of the parenting—not the parents' lifestyle—that matters most to kids. Sexual orientation alone doesn't make a person a good or bad parent. In Maplewood, N.J., Charlie and Marc are raising 17-month-old Olivia, whom they adopted. Last Christmas she had a lead role in their church's holiday pageant. "So you had a little Chinese girl of two gay parents who was the baby Jesus," says Charlie. Adds Marc: "It gives a whole new meaning to the word 'Mary'." As she gets older, Charlie and Marc say, they'll explain to Olivia why her family is unusual. "I think Olivia is so lucky to have the opportunity to be different," says Marc. "And that's what I intend to teach her."

Reflections

1. Do you know any families with lesbian or gay parents? What have been their experiences?

2. What are some ways in which the life cycle patterns of the families described in the article might differ from those of families with heterosexual parents? How might their patterns be similar to those of stepfamilies or adoptive families?

3. How could schools help foster understanding of gay families?

4. How can you help?

19

One of the most important tasks a newly married couple (especially a young couple) must perform is to establish their own family separate from the families of their parents. Jean Marzollo describes three young couples faced with the task of becoming independent.

Breaking Parental Ties

By Jean Marzollo

Ellen holds the set of calligraphy pens tightly on her lap and eyes the sleet hitting the windshield. "Please slow down," she says tensely.

"I'm not speeding," Tom answers, equally on edge. They are driving home from Ellen's mother's house, where they have been to a disastrous birthday part for Ellen. Although they've been married for only six months, they don't feel like newlyweds.

"I really do like the pens," says Ellen, who has recently taken up calligraphy.

"I felt so stupid when you were opening them," Tom replies. "You mother and father gave you so many presents, and they know I can't afford to give you much. I told you I didn't want to go there. Why couldn't we have gone out to dinner alone?"

"You know how Mother is," Ellen says. "She lives for our visits. And she loves birthdays. It would have hurt her terribly if we hadn't shown up."

"Well, I hate having to play second fiddle. I mean—for Pete's sake, your mother gave you a carload of expensive things and I gave you three felt-tipped calligraphy pens."

"No one cared. Why did you have to ruin the whole party with your sulking?"

Ellen begins to cry. She feels sorry for herself, it's her birthday after all—and sorry for her mother, who only wanted the evening to be festive. Ellen has always been her mother's pride and joy, and she feels it would be selfish not to visit as often as possible. She wonders why Tom can't understand.

Julie often feels just as Ellen does.

Julie's father, retired and recently widowed, lives alone in a town in Florida. She lives an hour away with her husband Wayne and their two-year-old son, Wayne Jr. Every day after Wayne leaves for work, Julie puts Wayne Jr. into his car seat and drives the forty miles to her father's house. There, her father and the baby watch TV while Julie cleans the house, weeds the garden, shops at the local supermarket and cooks a hearty, nutritious lunch with enough leftovers for her dad's dinner.

Often, Julie returns home too exhausted to shop and cook again, so when her husband comes home, she frequently suggests going to McDonald's. Wayne doesn't mind a burger and fries once in a while, but he misses Julie's cooking. He's beginning to feel angry that Julie cooks every day for her father and rarely for him. Their dinner hour, once a special time, is now hurried and harried.

Wayne's afraid to tell Julie what's really on his mind because he knows how much Julie loves her father and he's ashamed of feeling jealous of a lonely old man. He's afraid, too, that if he complains, Julie will suggest that her father move in with them. Wayne wants to avoid that because whenever Julie's dad is around, Wayne feels left out.

For her part, Julie's not happy about devoting more time and energy to her father than to her husband, but she believes that her dad really needs her. She's sure that if she stopped seeing her father so often, there would be a hole in his life that no one else could fill.

Ellen and Julie are having trouble with their marriages because they are too close to their parents. To complicate matters, neither woman realizes that the parents are the problem—both believe that the fault lies with their husbands.

© 1985 by Jean Marzollo, "Succeeding in the Union," *The San Jose Mercury News*

A great many young wives find themselves similarly torn between their husbands and their parents. Even though they're married, they haven't yet left the nest. According to Dr. Bruce M. Forester, assistant professor of psychiatry at Columbia University and a psychiatrist in private practice in New York City, "Overinvolvement with parents causes the most trouble in early marriage—more than sex or any other issue."

If you sense such a conflict in your marriage, keep these questions in mind:
- Who has the most control over your action—you, your parents or your husband?
- Whom do you care most about pleasing—yourself, your parents or your husband?
- Who depends on you most—your parents or your husband?

If your most frequent answer is "my parents," you do indeed have a problem. Being so involved with Mom and Dad endangers your marriage because it creates an imbalance; your life is weighted too heavily in your parents' favor. The solution isn't to shift all the weight over to your husband; instead, you need to put yourself in charge. You are the core of your own life, and you're now establishing the most intimate bond you'll ever experience—the one with your mate. That bond requires most of your attention right now because you're laying the foundation for a lifetime partnership. To be successful, you must let go of your parents.

Dr. Sonya Rhodes, a New York City family therapist and coauthor of *Surviving Family Life*...,, has written about stages in family development. She says you can be shackled to your parents by some very powerful bonds.

If you need your parents' money, for example, you may feel emotionally indebted to an excessive degree. Or if your parents are very attached to you and have centered their lives around you, you may feel extremely guilty about going your own way. You may fear that you will make your parents angry, sad—even sick—if you "abandon" them.

The first step, according to Forester, is to talk to your husband. How is he feeling—left out? Jealous? Hurt? Talk about your own feelings, which might include guilt, fear and confusion.

Once you've established that you do have a problem, "The next step is to draw up a plan together," says Forester. "You must work out a strategy that will gradually loosen the parent tie, but in a very tactful way."

If you're lucky enough to be able to talk directly to your parents, do so. For example, Julie might say to her father, "I love you very much, but I feel that visiting you every day is hurting my marriage. You're in good health and sociable. Maybe it's time you went out and met more people. Seeing less of each other will be better for both of us. From now on, I'll just come once a week."

Forester suggests that if you don't feel you can be blunt without causing tremendous hurt, you can back off more subtly.

For example, says Forester, if Julie couldn't speak frankly to her father, she might try withdrawing gradually. "In the first week, Julie can say, 'Look, Dad, I won't be able to come this Thursday.' And she can continue to cancel once a week for a few weeks. Then she can start canceling twice a week, always giving advance notice. Finally she can say, 'You know, it'd be great if you started seeing friends more often. I care about you and I want you to have companionship.' This way, everyone knows what's going on but no one needs to feel hurt or rejected."

Remember too that although making the break is very tough for both you and your parents, this crisis can bring you and your husband closer. It can reinforce your mutual commitment and strengthen your confidence in your ability to weather future crises together. Breaking away from your parents is almost guaranteed to be very hard, but the eventual rewards for everyone involved are immeasurable.

It isn't only wives who have trouble breaking away from Mom and Dad. Husbands can be overinvolved with parents too.

Gwen and Mark, for example, have been married a year. They don't have children yet, but Gwen feels they're already a family of three: Gwen, Mark, and Mark's mother. Mark's mother, a widow, is in the habit of dropping by a couple of times a week with homemade goodies. Naturally, she likes to chat awhile.

Mark doesn't mind his mother's frequent visits, but Gwen does. She has asked Mark to explain to his mother that the two of them, who both work, need more time alone. Lately, she hears herself nagging, and that frightens her. Mark feels torn between his mother and his wife. He wishes Gwen would understand his mother's good intentions and be nicer to her.

If you, like Gwen, are married to a man who's tied to his parents, you will have to tread carefully.
- Approach him gently. Be honest with your husband about the problems, but talk to him in a way that shows how much you care. Dr. Bruce M. Forester recommends saying

something like, "I know you love your parents very much, and I love them too, but I think we're spending too much time with them. I'd like to be alone with you more often. How can we work this out?"
- Avoid, at all costs, hurling accusations at your husband or his parents. "If you scold, you're acting like a mother," Dr. Forester says. "If his real mother is apt to yell at him too, he'll feel that everyone is treating him like a kid."
- If you resent visiting his parents often, Dr. Penelope Russianoff suggests your husband go alone occasionally. "You can say, 'I have a feeling your parents might enjoy being alone with you sometimes. Why don't I visit some friends while you go to your folks' this Sunday?' Afterward you can tell him what a great time you had," Russianoff adds. "He may feel more inclined to join you next time."
- You can't change his relationship with his parents overnight. It helps if you can see that his parents are probably well-meaning; they're not deliberately trying to interfere.

Reflections

1. What are some of the problems encountered in forming a separate family unit?

2. Do men and women experience this problem differently? Why?

3. Have you experienced similar problems to those described in the reading?

4. What are some ways in which young couples can resolve conflicting demands between their relationship and their parents?

20

Grandparents have a special role in the lives of their grandchildren. Anne McCarroll explains how her grandmother's kitchen served as a family gathering place where all members of the family could share their experiences and support each other.

At Grandma's Table

By Anne McCarroll

The happiest times of my childhood were not family outings, vacations, parties or recess. The happiest times were the hours spent around the kitchen table of my grandmother's house on Highland Avenue in Dayton, Ohio.

It was there the aunts, uncles and cousins gathered and talked for hours. They didn't all live there, of course, but they dropped in at their pleasure, sure of a warm welcome and something good to eat no matter the time of day.

No one ever, as I recall, played a game at that table. They talked and laughed. To me, it was more enjoyable than Monopoly, checkers or gin rummy. They didn't need to play games to be interested or entertained. Each came equipped with an active mind and a lively sense of humor.

There are families whose time together is ever structured around an activity—a game of Monopoly, tennis, sailing or cards—and I've sometimes felt sorry for them. There's a danger they'll miss the real fun—the talking, the sharing of thoughts and ideas, hopes and concerns, the funny things that happen, the characters they meet, the ironies of life.

At my grandmother's table there would be talk over dinner—and as people lingered over dessert. Then we would move to the living room—or, on warm summer evenings, to the backyard—for more conversation. Late in the evening, everyone would return to the table for another piece of pie, a dish of ice cream, a glass of milk.

Recently, reading Russell Baker's biography, *Growing Up*, I found that he, too, cherishes memories of kitchen table talk.

"I loved the sense of family warmth," Baker writes, "that radiated through long kitchen nights of talk.... I was receiving an education in the world and how to think about it. What I absorbed most deeply was not information but attitudes, ways of looking at the world that were to stay with me for many years."

I understand that because I still draw on deposits made in my memory during the hours around that table. No one preached or pontificated—as so many adults in the presence of children are apt to do. And yet, I learned so much about life.

I knew how each of my aunts and uncles felt about their jobs, their friends, religion, the state of the world, the man in the White House. And I knew their hopes and dreams.

There was not, around the table, a need to impress, a need to seem more than one was. They spoke of their failures as openly as their successes, of a mistake in judgment, a poor decision, a thoughtless remark or an embarrassing moment. I learned from their honesty, their openness, empathy and insights.

And they knew what I thought, what I felt—for we children entered into conversations. Our opinions were considered, our ideas respected.

The people around the table were different from one another. (Grammy did not turn out a uniform product.) It was not sameness that bound them together, but genuine caring for one another—and a shared sense of humor.

They knew they were not alike—talked about it, laughed about it. I suppose it was there that I learned not to tolerate differences—but to enjoy them.

I think this is why, although many people go through life always seeking their own kind, I've always preferred to know a variety of people—of different ages and interests. And I still enjoy the company of people older than I am as well as my peers.

It was at that kitchen table that I realized there is no right way to live, to think, to be. Grammy imposed no family norm to which one must conform to be accepted, to be loved.

My children never sat at Grammy's table. She was gone before they were born. But their lives are infinitely richer for all that I learned there.

And the joy that I've found in my own family was influenced enormously by the joy I found in the family of my childhood.

Families who find pleasure in conversation may stay closer, I think, than those in which the focus is on activities. One can play tennis or golf or bridge or backgammon with anyone who has learned the game. A golf or bridge partner is easily replaced. There is an ample supply of tennis partners—one needn't leave the neighborhood to find one. It is the conversations that are special—that make one's family unique and irreplaceable.

I want to know, when my cousin visits, how his life is going, not if his backhand has improved. (I can talk backhands with the neighbors.) I want to know which dreams have come true, which have been replaced by new ones.

But that's because of the hours at Grammy's table—where everyone cared more about what you were and what you thought—than what you did or where you went.

An acquaintance said recently, "My grandmother had a beautiful carved oak credenza; it was the most important piece of furniture in her house."

Grammy had some carved oak pieces, too. I never thought them important.

But her kitchen table—that was important.

Reflections

1. What values did Anne McCarroll learn at her grandmother's table?

2. What is (or was) your relationship with your grandparents? What did you learn from them?

3. Is there a place where your family typically gathers to share their thoughts and feelings? If not, why not?

4. Do you feel that society today allows families such opportunities to gather? Why or why not?

5. If you had (or have) grandchildren, how would you like your relationship to be?

As you read this essay, think about what it has to say about society's response to childless people. Think about your own reasons for wanting—or not wanting—children.

On Not Having Kids

By Jon Hubner

Before we left Boston for San Jose last fall, we had dinner with two couples we've known for years. We used to spend hours with them, discussing the pros and cons of starting a family. Within a few months, both couples had children. We thought it was going to be a bittersweet goodbye dinner, but they ganged up on us about having children as soon as we got our coats off. They kept telling us we didn't know what we were missing. We suggested they were missing a lot, too. That, they didn't like. "Well, people are having children later these days," one proud mother said. "There's still time. It's a matter of maturity. When you're ready, you'll know it"

That, I didn't like. Why should maturity be equated with having children? Before I could react, they hit us rapid-fire with a series of questions that probed like a scalpel. One new father, a psychiatrist, wondered if we might be just a bit narcissistic. His wife, a clinical psychologist, asked if we were perennial adolescents, afraid of the responsibility that comes with parenthood. She suggested that our relationship was too fragile for the strains of raising kids. They were as obnoxious as est graduates. We had two choices: laugh it off or tell them off.

"You want the truth?" I said. "Our sex life is so kinky, we'll never reproduce."

Nobody laughed. It was as if I'd committed heresy, joking about the hallowed act of creation. After a pause, the conversation shifted to a safer topic: California.

Ann and I are not narcissistic. Our relationship has endured worse strains than the "terrible twos." It's not that we do not like children, either. We have loved several. We have friends here who have a wonderful 2-year-old daughter, so I catch glimpses of what I'm missing. The point is, I've chosen to miss it. We do not have children because we do not want any. If we did, we would have them.

I have no deep biological urge to reproduce myself. The world will get along fine without my kids, just as it will get along fine without me when I'm gone. I lack the drive to love, to nurture, and to shape another human being. Babies don't fascinate me. I'm like Queen Victoria: I like kids after they've left what she called the "frog stage."

I worry about money like most of us do, but if we had kids, I'd *really* worry about money. I don't want to spend money on Gerber's and Pampers and life insurance. I don't want to save for a college education. I'd rather save so I can go back to Morocco. I don't want to spend nights babysitting. I want to go to country bars. I don't want to spend Saturdays taking 5-year-olds to Great America. I want to go to Reno. If this sounds hedonistic, perhaps it is. But it is not self-destructive. Aristotle, a really nifty definition-maker, said that to be happy is to be always learning something. My job does that for me. I've chosen not to learn about the things a child could teach me.

I'm not convinced I would love my child simply because he or she was mine. Plenty of parents do not like their kids. You've seen them. We have a great friend who is a successful novelist. She has reached her 60s, has no children, and has no regrets. "It's a real longshot to produce a child that you would choose as a friend," she says. "Think about your family: Your child could turn out to be like any one of them. How many of them do you like enough to want to spend the next 18 years with?

I suspect that in many cases, it is the parents' fault if they don't like their children. We had friends we stopped seeing because they have a kid who is a monster. The terror I'll call Joey is out of control the second his feet hit the floor. He is the product of what I call child idolatry: the child as center of the universe. The little creep controls every situation. Joey's parents are little more than serfs, the loyal subjects of a 4-year-old despot. "No, Joey, daddy can't read you a story now, daddy is busy talking . . . Joey, don't shout at daddy . . . Please don't shout, Joey . . . Joey, don't cry . . . Oh, all right Joey! Get the damn book."

I think that if I had a child, I would spend most of the day worrying about him. Jeffrey is the closest I've come to being a father; he and his mother moved in next door to us when he was 2. He was so beautiful, his hands and cute little legs emerging from his favorite Levi cutoffs so delicate, that I wanted to pick him up and hug him and never let him go. Every time he laughed, 1,000 volts shot up my spine. I'd think, do that again Jeffrey and I'll give you my house. If your best friend is the person you most enjoy being with, then that's what Jeffrey was to me. I liked the things he said and the way he said them. I liked watching *Sesame Street* with him. I liked the way he conned me into stopping for ice cream cones.

When Jeffrey was 6, they discovered he had cancer. His doctor told me he had a 50-50 chance to live six months. They wouldn't even give odds on a year. I took long walks, hunting the catharsis, trying to cry. Once I screamed and kicked a tree, but it was self-conscious and didn't help. Jeffrey was going to die.

He went through radiation and chemotherapy. When he lost his hair, his mother bought him a Dodger hat. He went back to school and never took the hat off. Any kid who tried to take it off was in for a fight. I refused to let myself hope, even when the cancer mysteriously went into remission.

Jeffrey survived! It is the one miracle I've seen. The cancer gradually disappeared; the doctors can't explain it. Jeffrey is 13 now, and he's still my man. But I never, ever, want to go through anything like that again.

Sometimes I think about moving into middle age, and then old age, without children. It is frightening. Loneliness is the enemy. For three years, we lived in New Hampshire, surrounded by retired couples. They wouldn't have had much to do besides read the Boston Herald American and bowl candlepins if it weren't for their children. Their families were scattered all over the country, so they didn't see their children or grandchildren often, but their families gave them a lot to think about. A letter or phone call was a major event. Still, with children or without, life for the elderly in this country is pretty bleak. We isolate old people. Those I've known spend far too much time alone. I'm hoping that if I make it to my 70s, there will be old-folks communes by then that will let me in.

As resolved as I am to not having children, I will always wonder what I have missed. The other day, I was walking around Vasona Park. I noticed a man who apparently had arrived late for a picnic. He was walking up a slight hill when his kids saw him. They jumped away from the picnic table and went running down the hill. The father scooped one up and swung him 'round and 'round while the other child danced about, waiting for an opening. Finally, the kid charged in and grabbed his father's leg. They wrestled for a while, and then walked hand-in-hand up to the picnic table.

As I watched, I wondered if the love that father has for his children is more profound than any love I've known. I'll always wonder about that.

Reflections

1. What are the author's reasons for not choosing to have children?

2. What is your response to his reasoning?

3. Do you think a person's decision not to have children needs to be justified to others?

4. In your own experience, are people who choose not to have children treated differently than those who have them?

5. What are your reasons for wanting—or not wanting—children?

In this family, both parents are equally responsible for caring for their baby. As you read this essay, think about what your own responses would be in a similar situation.

Sharing the Baby

By Joanne Kates

Thirteen months ago Leon was not interested in babies. They didn't speak the language, and he was sure that if he touched one it would cry.

But I refused to play Mom to his traditional Dad. The deal I had spent years fighting for—and won—was that each of us would take equal responsibility for our baby, which meant equal career sacrifices. Mara was born in August 1985. Leon had quit his job in July to get ready to be a full-time father for her first year. We knew that his career standing as a labor economist was good enough that he would have no trouble getting a new job when the time came. That was not the big risk. The risk was more personal: Would he like being a full-time parent? Without the validation he so frequently got in his work life, would he be able to believe in himself? He was not going to turn into an isolated drudge, because I would be working part time and therefore be with him and Mara most days. He wasn't going to lose his manhood because it was not solely vested in success. And yet the question remained. If he was not to be his usually hard-driving, overachieving man in a hurry, who would Leon be? Would it be enough for him? And for me? Our friends and families looked on with mixed emotions: they admired his guts, they envied his escape, however, temporary, from the rat race, and yet it worried them to see a successful man become a professional diaper-changer.

When Mara was a week old, on a sunny summer day, Leon and I stood in our driveway saying goodbye to visiting friends. He held the baby in one hand; the other hand was in his pocket. He was a model of nonchalance. He learned to take care of a baby the same way he learned public speaking, the same way a horse takes a fence: you see the challenge, you fear it and then you meet it. He had quickly found out that what taking care of Mara asked of him, he delighted to give: tenderness, a gentle touch, hugs by the dozen. She rewarded him by lying around looking beatific and Buddha-like. And he began to believe in himself as a nurturer.

When Mara was 3 months old, she got a cold. She woke frequently at night, crying because she couldn't breathe. The first time she awoke so miserable, Leon cried from sorrow and pity, his identification with her was so complete. It was a new definition of intimacy.

When Mara was 5 months old, we fought the war of the baby authorities. Leon wanted to feed her solid food for the first time. She was still being breast fed and I was not ready for a change. I cited Dr. Spock; he cited Penelope Leach. But the issue was not when to introduce baby to real food. The issue was *turf*. I clung to my exclusive connection. I, the staunch feminist who never wanted biology to determine destiny, was enjoying the responsibility of being Mara's sole source of nourishment. She had been fed by my body since she was smaller than a kernel of corn—for 14 months at that point. I got her to that smiling robust moment and I was possessive. It was a physical want; it underlay all our words of negotiation.

Leon wanted her that way too, and the only way he could have the satisfaction of feeding her was by starting her on solid food. I bought time with creative procrastination: after we get back from skiing...after we find the right baby cereal ....Just before she turned 6 months old, I gave in.

I am sometimes jealous of how much Leon dotes on Mara. When he calls her pet names previously spoken only to me, when he ignores me and pays attention to her, I am sometimes sad at being displaced. Their tie is blood, and mine with Leon is based only on volition. Ours can be broken; theirs cannot. Although their love delights me, there are times when she looks up at him adoringly and I am jealous. That is the price I pay for sharing her equally with him. Women who do all the parenting, or who share it with paid helpers, have a special power that I have relinquished. It is the power of being everything to a child. I am not Mara's everything. She cares as much for Leon as for me. This is most difficult when I walk up the stairs to my study to write, and leave the two of them playing in the living room. He blows me a kiss. She doesn't even look up.

When she was 8 months old, a grinning little imp in overalls, we had an adventure. She and I flew across the Atlantic together without Leon. It would have been fun but for the timing: it was a peak moment in terrorism in Europe, and Leon lost sleep for fear of something happening to us. He feared most for Mara, because she seemed so vulnerable. It came as a shock to him that he could not, try as he might, guarantee her a safe world. That journey stood for him as a symbol of what he could not do for her, and yet wanted so badly: he wanted to shield her from pain, to make sure she would grow up safe, healthy and happy. It was a painful lesson.

Summer came. We made our annual migration to the lake, where we live on an island and our main mode of transit is a canoe. We have neither electricity, telephone nor running water there. It was not designed with babies in mind. Mara was 11 months old, a competent crawler and indefatigable investigator of the world below the two-foot level. All summer he put socks and shoes on her and I took them off. I would set her down in the dirt to play. He would pick her up. I accused him of overprotecting her. He called me cavalier. When you throw the old roles into the mixer and turn it on, more changes result than the ones you negotiated. We had a full-scale role reversal on our hands. Leon the nervous mother, me the laissez-faire father. It was understandable why each of us was overplaying our roles somewhat, since they were so new to us and we had so few models to follow.

The week after Mara's first birthday an old friend of Leon's tracked us down in the woods. He wanted to offer Leon a senior government job, the kind only a fool refuses. It was a high-pressure job in another city, a dream job for a man on the fast track. Two years ago Leon would have tried hard to persuade me to move West for this. He sat me down after Mara was in bed, laughed ruefully and said, "Guess what I just turned down." His year with Mara has made Leon unfit for workaholism, because he's unwilling to settle for being no more than the bestower of a good-night kiss at the end of her day. I have been similarly altered by this child. So we have a new deal now: Leon will do consulting three days a week; I will write three days a week. The rest of the time is for Mara. Look for us at the zoo; it's less crowded on weekdays.

Reflections

1. What unexpected problems did the author encounter in sharing responsibility for rearing her daughter?

2. What concerns did the father have? What benefits did he gain from the arrangement?

3. Do you think such arrangements of shared parenting are possible for most families?

4. What kind of childrearing arrangement would be ideal for you? What would you be willing to give up in order to achieve it?

23

In this essay, the author writes of some of the unexpected consequences of deferred parenthood. As more and more people postpone having children until their thirties (or even forties), these issues will become increasingly commonplace.

The Lament of the Older Parent

By Laurie DeVault

I was 41 years old and just weeks away from giving birth to my first child. Three thousand miles away, my father lay dying of cancer, fearing he would never meet his grandchild. I secretly begged for him to hold on to life long enough for me to jump on a plane and rush to his side with my newborn son, but it was not to be.

The men and women of my generation are facing an unfortunate consequence of older parenthood: Our parents have begun to age and die before they have time to grandparent our children. Our Baby Boom generation grew up in the fifties and sixties, raised by parents who had married and begun parenting while still in their twenties. This meant there were plenty of grandparents around, youthful and energetic couples eager and able to be involved in their grandchildren's lives.

I had my beloved grandmother Mimi until I was in my thirties. She was without doubt the most significant person of my youth. Her cottage on the hill was a haven for me, a place where I always felt supremely special. On weekend visits I would be indulged with cookies and cinnamon rolls and enjoyed free access to her jewelry box and button collection. At night she would read me Margaret Wise-Brown's *Wait 'Til the Moon Is Full*, and mornings I'd wake up to Swedish pancakes or waffles, cantaloupe and strawberries.

Mimi indulged me in another way, a way that seems to be the particular domain of grandparents: she gave me the unconditional love and acceptance that every child needs. My parents were, well, parents, and as such were embroiled in the daily struggles of rearing their children to be tidy, responsible, well-educated, polite, and productive citizens. As a result, our home was like thousands of others across the country, filled with fights over allowances and dirty dishes, resentment over being nagged about homework, and tension and despair during those painful times when children can feel so misunderstood. At Mimi's house I could always count on being understood. It was there I was listened to and comforted and my injured ego soothed. In her calm and loving presence I felt worthwhile and lovable. I felt "good" and wanted "be" good.

Like many other grandparents, Mimi gave me another gift: A sense of heritage. One of the greatest pleasures of my visits was hearing her stories about the "olden days." I learned Swedish nursery rhymes and listened to stories about Mimi's gambling lumberjack brother, her poetry-writing mother, and her gentle fiddle-playing father. I learned about making bread, soap, butter and blackberry jam, and about keeping the floor clean with a broom and damp wads of newspaper.

Mimi also told me about love. When she would hold my hand and tell me of the life she shared with my grandfather, her eyes would fill with the tears of tender memories. Papa was much older than she and died when she was only 60. When she told me how it felt to have such love, I knew I wanted the same thing for myself.

I did find such a love, but not until I was 40. By that time Mimi had been gone for six years. My husband and child would never know her. I thought that if I couldn't share my grandparents with my children, I most certainly could count on sharing my parents with them. I always assumed my children would have the same pleasure of grandparents that I had. It never occurred to me how my delayed parenthood would prevent this.

Now with both my father and a beloved stepfather gone, and my mother not so young, I worry that my child will never have his own haven away from family turmoils, a place where he can be indulged and feel supremely special. I feel sad that as he grows older he may never know the comfort of a plump, grandmotherly hug or the joy of gazing into twinkling aged eyes while hearing stories of the olden days.

Another sense of loss stems from knowing I may never get a chance to be that plump twinkling grandmother myself. If my child does have children, he may delay parenthood himself, rendering me one very aged (or dead) grandma.

There is hope, however. If my spirited mother can stick around for a while (as she's promising) my son will be able to get in some quality grandmother time, even though we're on opposite coasts. We're also lucky to have my husband's parents, who live closer and play a significant role in our son's life. But they aren't so young, either, and I wonder just how long they'll be physically and emotionally available to their grandson.

I also find hope in my plan to make sure my son knows his late grandfather. I'm putting together an album all about my father containing letters, photographs and other mementos. I urge other parents to do the same. If your parents are still alive, consider videotaping them. I wish I had, and plan to do so with my son's living grandparents.

Finally, I must grieve for my son's siblings that are never to be. As one who feels deeply for her brother and sister, I am saddened by the sinking reality that I cannot provide the same for my own child because I have grown too old. He will have no lifetime peer to grow up with, no aunts or uncles (from his side of the family) for his own children, and no one to share the burden of caring for his aging parents.

So I now understand why people have traditionally become parents when they're young. And though I may lament that I waited so long, I know I simply wasn't ready. I had a lot of growing up to do. I had a career to get started, relationships to work out and countries to explore. I also had to meet my husband. He, and the little boy I hold in my arms, have made the wait worthwhile and the losses easier to bear.

Reflections

1. What are some of the losses that the author faces as a result of deferring parenthood?

2. Do you know anyone who has deferred parenthood until their thirties or forties? If so, have they faced issues similar to those described by Ms. DeVault? Have they faced other issues not discussed in the reading?

3. Ideally, at what age would you like to have a child? What are the pros and cons of parenting at that age?

24

This article by sociologist Arlie Hochschild is excerpted from her book The Time Bind: When Work Becomes Home and Home Becomes Work. *Here she explores the tension between paid work outside the home and family/household work. As you read the article, think about the non-monetary advantages of employment, especially as they apply to women.*

There's No Place Like Work

Excerpted from *The Time Bind*
By Arlie Russell Hochschild

It's 7:40 A.M. when Cassie Bell, 4, arrives at the Spotted Deer Child-Care Center, her hair half-combed, a blanket in one hand, a fudge bar in the other. "I'm late," her mother, Gwen, a sturdy young woman whose short-cropped hair frames a pleasant face, explains to the child-care worker in charge. "Cassie wanted the fudge bar so bad, I gave it to her," she add apologetically.

"*Pleeese*, can't you take me with you?" Cassie pleads.

"You know I can't take you to work," Gwen replies in a tone that suggests that she has been expecting this request. Cassie's shoulders droop. But she has struck a hard bargain—the morning fudge bar—aware of her mother's anxiety about the long day that lies ahead at the center. As Gwen explains later, she continually feels that she owes Cassie more time than she gives her—she has a "time debt."

Arriving at her office just before 8, Gwen finds on her desk a cup of coffee in her personal mug, milk no sugar (exactly as she likes it), prepared by a co-worker who managed to get in ahead of her. As the assistant to the head of public relations at a company I will call Amerco, Gwen has to handle responses to any reports that may appear about the company in the press—a challenging job, but one that gives her satisfaction. As she prepares for her first meeting of the day, she misses her daughter, but she also feels relief; there's a lot to get done at Amerco.

Gwen used to work a straight eight-hour day. But over the last three years, her workday has gradually stretched to eight and a half or nine hours, not counting the E-mail messages and faxes she answers from home. She complains about her hours to her co-workers and listens to their complaints—but she loves her job. Gwen picks up Cassie at 5:45 and gives her a long, affectionate hug.

At home, Gwen's husband, John, a computer programmer, plays with their daughter while Gwen prepares dinner. To protect the dinner "hour"—8:00-8:30—Gwen checks that the phone machine is on, hears the phone ring during dinner but resists the urge to answer. After Cassie's bath, Gwen and Cassie have "quality time," or "Q.T.," as John affectionately calls it. Half an hour later, at 9:30, Gwen tucks Cassie into bed.

There are, in a sense, two Bell households: the rushed family they actually are and the relaxed family they imagine they might be if only they had time. Gwen and John complain that they are in a time bind. What they say they want seems so modest—time to throw a ball, to read to Cassie, to witness the small dramas of her development, not to speak of having a little fun and romance themselves. Yet even these modest wishes seem strangely out of reach. Before going to bed, Gwen has to E-mail messages to her colleagues in preparation for the next day's meeting; John goes to bed early, exhausted—he's out the door by 7 every morning.

Nationwide, many working parents are in the same boat. More mothers of small children than ever now work outside the home. In 1993, 56 percent of women with children between 6 and 17 worked outside the home full time year round; 43 percent of women with children 6 and under did the same. Meanwhile, fathers of small children are

---

From: *The Time Bind: When Work Becomes Home and Home Becomes Work* by Arlie Russell Hochschild, © 1997 by Arlie Russell Hochschild. Reprinted by permission of Henry Holt and Company, Inc.

not cutting back hours of work to help out at home. If anything, they have increased their hours at work. According to a 1993 national survey conducted by the Families and Work Institute in New York, American men average 48.8 hours of work a week, and women 41.7 hours, including overtime and commuting. All in all, more women are on the economic train, and for many—men and women alike—that train is going faster.

But Amerco has "family friendly" policies. If your division head and supervisor agree, you can work part time, share a job with another worker, work some hours at home, take parental leave or use "flex time." But hardly anyone uses these policies. In seven years, only two Amerco fathers have taken formal parental leave. Fewer that 1 percent have taken advantage of the opportunity to work part time. Of all such policies, only flex time—which rearranges but does not shorten work time—has had a significant number of takers (perhaps a third of working parents at Amerco).

Forgoing family-friendly policies is not exclusive to Amerco workers. A 1991 study of 188 companies conducted by the Families and Work Institute found that while a majority offered part-time shifts, fewer than 5 percent of employees made use of them. Thirty-five percent offered "flex place"—work from home—and fewer than 3 percent of their employees took advantage of it. And an earlier Bureau of Labor Statistics survey asked workers whether they preferred a shorter workweek, a longer one or their present schedule. About 62 percent preferred their present schedule; 28 percent would have preferred longer hours. Fewer than 10 percent said they wanted a cut in hours.

Still, I found it hard to believe that people didn't protest their long hours at work. So I contacted Bright Horizons, a company that runs 136 company-based child-care centers associated with corporations, hospitals and Federal agencies in 25 states. Bright Horizons allowed me to add questions to a questionnaire they sent out to 3,000 parents whose children attended the centers. The respondents, mainly middle-class parents in their early 30's, largely confirmed the picture I'd found at Amerco. A third of fathers and a fifth of mothers described themselves as "workaholic," and 1 out of 3 said their partners were.

To be sure, some parents have tried to shorten their hours. Twenty-one percent of the nation's women voluntarily work part time, as do 7 percent of men. A number of others make under-the-table arrangements that don't show up on surveys. But while working parents say they need more time at home, the main story of their lives does not center on a struggle to get it. Why? Given the hours parents are working these days, why aren't they taking advantage of an opportunity to reduce their time at work?

The most widely held explanation is that working parents cannot afford to work shorter hours. Certainly this is true for many. But if money is the whole explanation, why would it be that at places like Amerco, the best-paid employees—upper-level managers and professionals—were the least interested in part-time work or job sharing, while clerical workers who earned less were more interested?

Similarly, if money were the answer, we would expect poorer new mothers to return to work more quickly after giving birth than rich mothers. But among working women nationwide, well-to-do new mothers are not much more likely to stay home after 13 weeks with a new baby than low-income new mothers. When asked what they look for in a job, only a third of respondents in a recent study said salary came first. Money is important, but by itself, money does not explain why many people don't want to cut back hours at work.

A second explanation goes that workers don't dare ask for time off because they are afraid it would make them vulnerable to layoffs. With recent downsizings at many large corporations, and with well-paying, secure jobs being replaced by lower-paying, insecure ones, it occurred to me that perhaps employees are "working scared." But when I asked Amerco employees whether they worked long hours for fear of getting on a layoff list, virtually everyone said no. Even among a particularly vulnerable group—factory workers who were laid off in the downturn of the early 1980's and were later rehired—most did not cite fear for their jobs as the only, or main, reason they worked overtime. For unionized workers, layoffs are assigned by seniority, and for nonunionized workers, layoffs are usually related to the profitability of the division a person works in, not to an individual work schedule.

Were workers uninformed about the company's family-friendly policies? No. Some even mentioned that they were proud to work for a company that offered such enlightened policies. Were rigid middle managers standing in the way of workers using these policies? Sometimes. But when I compared Amerco employees who worked for flexible managers with those who worked for rigid managers, I found that the flexible managers reported only a few more applicants than the rigid ones. The evidence, however counterintuitive, pointed to a paradox: workers at the company I studied weren't protesting the time bind. They were accommodating to it.

Why? I did not anticipate the conclusion I found myself coming to: namely, that work has become a form of "home" and home has become "work." The worlds of home and work have not begun to blur, as the conventional wisdom goes, but to reverse places. We are used to thinking that work is where most people feel like "just a number" or "a cog in a machine." It is where they have to be "on," have to "act," where they are least secure and most harried.

But new management techniques so pervasive in corporate life have helped transform the workplace into a more appreciative, personal sort of social world. Meanwhile, at home the divorce rate has risen, and the emotional demands have become more baffling and complex. In addition to teething, tantrums and the normal developments of growing children, the needs of elderly parents are creating more tasks for the modern family—as are the blending, unblending, reblending of new stepparents, stepchildren, exes and former in-laws.

This idea began to dawn on me during one of my first interviews with an Amerco worker. Linda Avery, a friendly, 38-year-old mother, is a shift supervisor at an Amerco plant. When I meet her in the factory's coffee-break room over a couple of Cokes, she is wearing blue jeans and a pink jersey, her hair pulled back in a long, blond ponytail. Linda's husband, Bill, is a technician in the same plant. By working different shifts, they manage to share the care of their 2-year-old son and Linda's 16-year-old daughter from a previous marriage. "Bill works the 7 A.M. to 3 P.M. shift while I watch the baby," she explains. "Then I work the 3 P.M. to 11 P.M. shift and he watches the baby. My daughter works at Walgreen's after school."

Linda is working overtime, and so I begin by asking whether Amerco required the overtime, or whether she volunteered for it. "Oh, I put in for it," she replies. I ask her whether, if finances and company policy permitted, she'd be interested in cutting back on the overtime. She takes off her safety glasses, rubs her face and, without answering my question, explains: "I get home, and the minute I turn the key, my daughter is right there. Granted, she needs somebody to talk to about her day. . . . The baby is still up. He should have been in bed two hours ago, and that upsets me. The dishes are piled in the sink. My daughter comes right up to the door and complains about anything her stepfather said or did, and she wants to talk about her job. My husband is in the other room hollering to my daughter, 'Tracy, I don't ever get any time to talk to your mother, because you're always monopolizing her time before I even get a chance!' They all come at me at once."

Linda's description of the urgency of demands and the unarbitrated quarrels that await her homecoming contrast with her account of arriving at her job as a shift supervisor: "I usually come to work early, just to get away from the house. When I arrive, people are there waiting. We sit, we talk, we joke. I let them know what's going on, who has to be where, what changes I've made for the shift that day. We sit and chitchat for 5 or 10 minutes. There's laughing, joking, fun.

For Linda, home has come to feel like work and work has come to feel a bit like home. Indeed, she feels she can get relief from the "work" of being at home only by going to the "home" of work. Why has her life at home come to seem like this? Linda explains it this way: "My husband's a great help watching our baby. But as far as doing housework or even taking the baby when I'm at home, no. He figures he works five days a week; he's not going to come home and clean. But he doesn't stop to think that I work seven days a week. Why should I have to come home and do the housework without help from anybody else? My husband and I have been through this over and over again. Even if he would just pick up from the kitchen table and stack the dishes for me, that would make a big difference. He does nothing. On his weekend off, he goes fishing. If I want any time off, I have to get a sitter. He'll help out if I'm not here, but the minute I am, all the work at home is mine."

With a light laugh, she continues: "So I take a lot of overtime. The more I get out of the house, the better I am. It's a terrible thing to say, but that's the way I feel."

When Bill feels the need for time off, to relax, to have fun, to feel free, he climbs in his truck and takes his free time without his family. Largely in response, Linda grabs what she also calls "free time"—at work. Neither Linda nor Bill Avery wants more time together at home, not as things are arranged now.

How do Linda and Bill Avery fit into the broader picture of American family and work life? Current research suggests that however hectic their lives, women who do paid work feel less depressed, think better of themselves and are more satisfied than women who stay at home. One study reported that women who work outside the home feel more valued at home than housewives do. Meanwhile, work is where many women feel like "good mothers." As Linda reflects: "I'm a good mom at home, but I'm a better mom at work. At home, I get into fights with Tracy. I want her to apply to a junior college, but she's not interested. At work, I think I'm better at seeing the other person's point of view."

Many workers feel more confident they could "get the job done" at work than at home. One study found that only 59 percent of workers feel their "performance" in the family is "good or unusually good," while 86 percent rank their performance on the job this way.

Forces at work and at home are simultaneously reinforcing this "reversal." The lure of work has been enhanced in recent years by the rise of company cultural engineering—in particular, the shift from Frederick Taylor's principles of scientific management to the Total Quality principles originally set out by W. Edwards Deming. Under the influence of a Taylorist world view, the manager's job was to coerce the worker's mind and body, not to appeal to the worker's heart. The Taylorized worker was de-skilled, replaceable and cheap, and as a consequence felt bored, demeaned and unappreciated.

Using modern participative management techniques, many companies now train workers to make their own work decisions, and then set before their newly "empowered" employees moral as well as financial incentives. At Amerco, the Total Quality worker is invited to feel recognized for job accomplishments. Amerco regularly strengthens the familylike ties of co-workers by holding "recognition ceremonies" honoring particular workers or self-managed production teams. Amerco employees speak of "belonging to the Amerco family," and proudly wear their "Total Quality" pins or "High Performance Team" T-shirts, symbols of their loyalty to the company and of its loyalty to them.

The company occasionally decorates a section of the factory and serves refreshments. The production teams, too, have regular get-togethers. In a New Age recasting of an old business slogan— "The Customer Is Always Right"— Amerco proposes that its workers "Value the Internal Customer." This means: Be as polite and considerate to co-workers inside the company as you would be to customers outside it. How many recognition ceremonies for competent performance are being offered at home? Who is valuing the internal customer there?

Amerco also tries to take on the role of a helpful relative with regard to employee problems at work and at home. The education-and-training division offers employees free courses (on company time) in "Dealing With Anger," "How to Give and Accept Criticism," " How to Cope With Difficult People."

At home, of course, people seldom receive anything like this much help on issues basic to family life. There, no courses are being offered on "Dealing With Your Child's Disappointment in You" or "How to Treat Your Spouse Like an Internal Customer."

If Total Quality calls for "re-skilling" the worker in an "enriched" job environment, technological developments have long been de-skilling parents at home. Over the centuries, store-bought goods have replaced homespun cloth, homemade soap and home-baked foods. Day care for children, retirement homes for the elderly, even psychotherapy are, in a way, commercial substitutes for jobs that a mother once did at home. Even family-generated entertainment has, to some extent, been replaced by television, video games and the VCR. I sometimes watched Amerco families sitting together after their dinners, mute but cozy, watching sitcoms in which television mothers, fathers and children related in an animated way to one another while the viewing family engaged in relational loafing.

The one "skill" still required of family members is the hardest one of all—the emotional work of forging, deepening or repairing family relationships. It takes time to develop this skill, and even then things can go awry. Family ties are complicated. People get hurt. Yet as broken homes become more common—and as the sense of belonging to a geographical community grows less and less secure in an age of mobility—the corporate world has created a sense of "neighborhood," of "feminine culture," of family at work. Life at work can be insecure; the company can fire workers. But workers aren't so secure at home, either. Many employees have been working for Amerco for 20 years but are on their second or third marriages or relationships. The shifting balance between these two "divorce rates" may be the most powerful reason why tired parents flee a world of unresolved quarrels and unwashed laundry for the orderliness, harmony and managed cheer of work. People are getting their "pink slips" at home.

Amerco workers have not only turned their offices into "home" and their homes into workplaces; many have also begun to "Taylorize" time at home, where families are succumbing to a cult of efficiency previously associated mainly with the office and factory. Meanwhile, work time, with its ever longer hours, has become more hospitable to sociability—periods of talking with friends on E-mail, patching up quarrels, gossiping. Within the long workday of many Amerco employees are great hidden pockets of inefficiency while, in the far smaller number of waking weekday hours at home, they are, despite themselves, forced to act increasingly time-conscious and efficient.

The Averys respond to their time bind at home by trying to value and protect "quality time." A concept unknown to their parents and grandparents, "quality time" has become a powerful symbol of the struggle against the growing pressures at home. It reflects the extent to which modern parents feel the flow of time to be running against them. The premise behind "quality time" is that the time we devote to relationships can somehow be separated from ordinary time. Relationships go on during quantity time, of course, but then we are only passively, not actively, wholeheartedly, specializing in our emotional ties. We aren't "on." Quality time at home becomes like an office appointment. You don't want to be caught "goofing off around the water cooler" when you are "at work."

Quality time holds out the hope that scheduling intense periods of togetherness can compensate for an overall loss of time in such a way that a relationship will suffer no loss of quality. But this is just another way of transferring the cult of efficiency from office to home. We must now get our relationships in good repair in less time. Instead of nine hours a day with a child, we declare ourselves capable of getting "the same result" with one intensely focused hour.

Parents now more commonly speak of time as if it is a threatened form of personal capital they have no choice but to manage and invest. What's new here is the spread into the home of a financial manager's attitude toward time. Working parents at Amerco owe what they think of as time debts at home. This is because they are, in a sense, inadvertently "Taylorizing" the house—speeding up the pace of home life as Taylor once tried to "scientifically" speed up the pace of factory life.

Reflections

1. What are the attractions of employment for the workers discussed in the article?

2. In what ways can the workplace be seen as fulfilling the functions of a family?

3. What are the drawbacks of full-time employment for the workers in the article?

4. According to Hochschild, why don't more workers take advantage of parental leave and other "family friendly" policies?

5. Do you believe both parents should work full-time? What would be the ideal situation for you?

25

Most families find themselves under constant pressure, torn between the needs of family and the demands of work. The following article offers some solutions for dealing with the time bind.

Beating the Clock

By Barbara Kantrowitz

Family-friendly employers and changes in public policy can help ease the household-stress overload, but individual ingenuity is still the critical survival skills for two-career families in the 1990s. "There's no such thing as routine parenting anymore," says Bennett L. Leventhal, a child psychiatrist at the University of Chicago and the father of three young children. "We have to be very creative, very innovative and very flexible in order to account for the kids' busy schedules and our busy schedules." Here are some suggestions:

Slow down: Families on the higher end of the income scale have the most options. One spouse can cut back on hours for a few years without too much financial sacrifice. (Remember, a reduced schedule also means savings on child care and commuting.) When her first child was born, Maureen Parton, a San Francisco litigator, switched to a job as an aide to a county supervisor because it offered more flexibility. Now she shares that position with two other women and often works only until mid-afternoon so she can pick up her kids, ages 6 and 3, and ferry them to ballet or soccer. Her husband, Jim, also a lawyer, often is home too late for dinner, but that hasn't stopped the family from eating one meal a day together. At 7:30 each morning, "we sit down and have 30 minutes for breakfast," says Parton. "You have to take the moments when you can get them."

Bargain for time: In the last decade, hundreds of businesses around the country have responded to the needs of working parents by offering alternatives like job sharing, flexible hours and telecommuting. But many employees, especially fathers, are reluctant to take advantage of these opportunities because they think that putting family before work—even with the boss's OK—will mean a permanent switch to the slow lane. Downsizing has only added to their anxiety.

While these concerns are legitimate, the workplace environment won't improve until employers perceive a "critical mass" of parents demanding change. That movement could start with employees who are secure in their standing, especially the office stars. They can bargain for benefits—for example, unpaid time off during school vacations or a reduced schedule (at a lower pay)—that enable them to be around for the important events in their kids' lives. The more experience employers have with staffers who have successfully balanced professional and personal lives, the more likely they are to extend these benefits to other workers.

Stay flexible: Sometimes parents find that it takes three or even more tries before they find the right answer. There's no one-size-fits-all solution. When she worked for a national trade association based in Florida, Jane Blackman, a 43-year-old single mother, struggled every morning to get her daughter, Holly, ready for day care or school. "There were tears all the way because she didn't want to say goodbye," says Blackman, whose position required her to be on call 24 hours a day. In 1991 Blackman took a non-profit job in Portland, Ore., because she thought the hours would be more manageable. But soon, she says, "my workdays were getting longer and longer . . . I was back in the rat race." Last year she finally struck out on her own, starting a business, Creative Possibilities, that arranges tea parties for local organizations and corporations. She's not making big bucks, but both mother and daughter say the sacrifice is worth it. A major plus: Blackman is always around when Holly, now 11, gets home

From *Newsweek*, 05/12/97 © 1997, Newsweek, Inc. All rights reserved. Reprinted by permission.

from school. Says Holly: "I like my mom just the way she is right now."

Be consistent: Steffan Kraehmer, author of "Quantity Time: Moving Beyond the Quality Time Myth," recommends that parents concentrate on "the three Rs of memory-making—routines, rituals and the ridiculous." The first two can start at mealtime: families should make every effort to eat together at least once a day—with the TV off. Those few minutes give everybody a chance to reconnect. Rituals, which could include everything from formal holiday celebrations to making Wednesday a regular pizza night, give kids the structure they need to feel secure in their identity as part of a family. Establishing these traditions provides kids and parents an opportunity to work together; everyone can be part of the decision-making process.

Celebrate everyday miracles: The "ridiculous" moments are pure serendipity, unscripted adventures that should be cherished and celebrated, because they provide the raw material for what ultimately becomes family legend. Like the time Kraehmer and his son Ryan, then 5, were walking through the orchard of their upstate—New York home. Ryan got stuck in the springtime mud, which he called "100 percent yuck." Six years later, it's still a favorite family story. It wasn't "quality" time in the classic sense. But you had to be there.

Reflections

1. As a child, parent, or other household member, have you experienced stress related to the multiple demands of family and work (or school)?

2. Do you (or did you) eat at least one meal a day with your family? Why or why not?

3. In your experience, does the workplace offer real alternatives for working parents?

4. Of the solutions suggested, which do you believe might work for you and why?

26

In the current debate about welfare, many argue that female and child poverty is related to family structure, that is, the single-parent family. They advocate marriage or remarriage for single mothers as the solution for the "feminization of poverty." The authors present an alternative view.

The Feminization of Poverty

By Marjorie E. Starrels, Sally Bould, and Leon J. Nicholas

Most analysts agree that low marriage rates and high divorce rates contribute to the feminization of poverty. If all parents were married, poverty would be more equally distributed between men and women. However, some suggest a stronger causal link. Novak (1987), for instance, posits that if a lasting marriage were the almost universal choice and if young persons postponed childbearing until they had completed school, married, and established themselves in adequate employment, then dependency would fall. Eggebeen and Lichter (1991) similarly attempt to demonstrate that changing patterns of family formation have resulted in increased child poverty. They claim that changes in Black family structure from 1960 to 1988 resulted in more poverty for Black children relative to White children. This kind of argument has resulted in Smith's (1988) conclusion that decreasing racial inequality and improving the economic status of Black children require changing marriage patterns more than changing employment patterns. Yet Stern (1993) provides evidence that a woman who divorces or separates from her chronically jobless husband and enters the labor force will hardly change her poverty risk at all.

Policies that advocate marriage for female householders are questionable from a practical as well as from a humanist and feminist perspective. Eggebeen and Lichter (1991), for instance, do not discuss the fact that many of the currently unmarried men who comprise the eligible marriage market for these women are unemployed and/or homeless. These nonincarcerated men would have difficulty supporting children above the poverty line. Therefore, urging women with children to get married is not a viable solution despite its popularity among the profamily lobby in the nation's capital. Further, given the positive relationship between male unemployment and family violence, pressuring poor and unemployed men into marriage could result in increased wife and child abuse rather than in decreased poverty.

White women have greater opportunities to escape poverty through remarriage because proportionately fewer White men are poor, homeless, unemployed, or incarcerated. These remarriages are typically accompanied by increases in family income. Caution is indicated, however, by children's lower rates of educational achievement in stepfamilies compared with those in two-parent families of a similar socioeconomic status. Adding a wage-earning stepfather may create other family problems even though it improves the financial standing of the mother and children (McLanahan et al., 1989).

The debate over marriage as a solution to children's poverty focuses disproportionately on Blacks (see Smith, 1988). Concern with Black marriage and sexuality has a long academic tradition in the United States. Yet if one considers other impoverished minority groups such as Mexican Americans, for which nearly two-thirds of poor children live in married-couple families, there is a clear need for a broader policy focus. Even the majority of poor White children live in married-couple families.

Despite the rapidly increasing proportion of Black children living in female-householder families from 1960 to 1980, poverty among Black children decreased dramatically from 66% in 1960 to 37% in 1980 (Eggebeen & Lichter, 1991). The only increase in Black and White children's poverty occurred between 1980 and 1988 and was probably influenced by the reduction in welfare benefits and wages. Welfare payments fell to levels significantly below the poverty level even in states such as New York that have had relatively high welfare payments.

Wages for unskilled and semiskilled workers also declined, especially among high school graduates. Blank (1991) demonstrates conclusively that falling wages during the 1980s, following a period of rising wages, was the single most important cause of increased poverty. Changes in family structure from 1980 to 1988 accounted for less than 5% of the increase in poverty.

Therefore, from a policy standpoint, the elimination of poverty is a more sensible goal than is the defeminization of poverty. The rates of child poverty in the United States are the highest in the industrialized world. (Smeeding, Torrey, & Rein, 1986). Although good jobs are critical for the economic self-sufficiency of both women and men, a reduction in children's poverty requires us to focus attention on women. If policy efforts are directed toward increasing marriage and remarriage rates, they are likely to fail or to increase wife and child abuse. On the other hand, if these efforts are directed toward providing child care, medical benefits, and jobs that provide a living wage, they have a greater chance of success.

It may be appropriate to move beyond the moralistic statements of Dan Quayle (Morrow, 1992) and the narrow statistical vision of "let's go back to the 1960s" family structure. There are many dedicated parents—both married and unmarried—who desperately need assistance in providing adequate food, shelter, medical care, and education for their children. These unmet needs, couched within their diverse social and familial contexts, may provide the stimulus for specific policy proposals with regard to job creation, health care reform, and expansions of child care.

Blank, R. M. (1991, July). *Why were poverty rates so high in the 1980s?* Working Paper No. 57, Jerome Levy Institute, Annandale-on-Hudson, NY.

Eggebeen, D. J. & Lichter, D. T. (1991). Race, family structure, and changing poverty among American children. *American Sociological Review*, 56, 801-817.

McLanahan, S. S., Sorensen, A., & Watson, D. (1989) Sex differences in poverty, 1950-1980. *Signs*, 15, 102-122.

Morrow, L. (1992, June 1). But seriously folks... *Time*, pp. 28-31.

Novak, M. (1987). *The new consensus on family and welfare: A community of self-reliance.* Washington, D.C.: American Enterprise Institute.

Smeeding, T., Torrey, B. B., & Rein, M. (1986, May). *The economic status of the young and old in six countries.* Paper presented at the American Association for the Advancement of Science annual meeting, Philadelphia.

Smith, J. P. (1988). Poverty and the family. In G. D. Sandefir & M. Tienda (Eds.) *Divided opportunities: Minorities, poverty, and social policy* (pp. 141-172). New York: Plenum.

Stern, M. J. (1993). Poverty and family composition since 1940. In B. M. Katz (Ed.), *The underclass debate* (pp. 220-253). Princeton, NJ: Princeton University Press.

Reflections

1. What are the values inherent in the argument that single parenthood is a cause of poverty? What opinions, stereotypes, and biases are demonstrated?

2. What reasons do the authors present for the increase in poverty among women and children?

3. What role does race or ethnicity play in the feminization of poverty?

4. What solutions or suggestions do the authors offer?

5. In your opinion, what government policies might address these issues?

27

Although the survey described in this article relates to working women, many of the findings apply to men as well. As you read, reflect on the stresses in your own life and on ways in which you might be able to ease some of the strain.

What Pushes Your Stress Button?

By Anne Cassidy

Work has become more stressful in the past two years, say more than three quarters (78 percent) of the working mothers in our most recent stress survey. In fact, the majority of respondents say an increased work load is their greatest job stress.

Amazingly, the demands of a complex, multirole life don't seem to have dampened working mothers' spirits. Despite the escalating pressure of the workplace, respondents indicate they are resilient and determined to succeed. The majority report they are happier than their mothers were. Moreover, they love their dual roles—only 11 percent say they would never work again even if they won the lottery!

These are some of the findings of an important survey that ran in spring issues of *Working Mother* and *Working Woman* magazines. Of the many thousands of replies, slightly more than 1,000 were tabulated, 563 of which were from working mothers. (The numbers used in this report are from the working mother sample, unless otherwise noted.)

As upbeat as women are about overcoming the stress in their lives, however, staying on top of it exacts a price. Many women are low on energy and short on time for themselves. Worst of all, the pressures they feel from work are straining their relationships with their loved ones, especially their mates. But today's working mothers know what would make their lives better: more flexibility, more money, more time for themselves and more help at home and on the job.

"What women are saying is 'I like being with my family and I like my work. I just wish there was a better way to manage them both,'" says Dana Friedman, co-president of the Families and Work Institute in New York City.

Our survey paints a vivid picture of women who are stretching their inner resources to make their lives work. Here are some of the most important findings:

- **Most women feel very stressed.** When asked to rate their overall stress on a scale of one to five (five begin the highest), almost half the respondents (47 percent) gave themselves a four.
- **Jobs are becoming increasingly demanding.** Sixty-eight percent describe their work load as either "uneven and unpredictable" or "too heavy to do a good job."
- **Stress exacts a toll on personal relationships.** Seventy percent of women report a strained relationship with either their spouse or children as the biggest intangible toll stress takes on their lives.
- **Despite these pressures, women feel they're on the right track.** Sixty-nine percent say they are happier than their mothers were. Being busy seems to have its own interesting and complex rewards.

Where Stress Comes From

The survey reflects the realities of an unpredictable economy: Working mothers are experiencing job uncertainty and the strain of handling extra work left behind by those who've lost their jobs.

"The stress I feel comes from needing to work 50-plus hours a week to make up for the lack of staff," writes Debbie Pence, a mother of two young children who works in a telecommunications company in Pennsylvania. "Before this insanity at work, I feel my stress level was a two on your scale, with most of the stress being positive," she says. Under these present circumstances, Pence rates her stress as "mostly negative."

Her plight is common. When asked to consider their greatest source of job stress, more than half the respondents—61 percent—picked an increased work load. This was almost twice as many as chose responsibility without authority (32 percent), a classic measure of stress. With so much work to shoulder, it's no wonder that 45 percent of women feel stressed out once or twice a week and 38 percent more than three times a week.

Other major sources of stress are the housework and child care women face when they come home. Sixty-one percent say they take on most of the responsibility for their kids, 26 percent share the work equally, and only 6 percent say their partners assume the responsibility for child care. On average, women report spending about an hour each weekday on household chores, with many explaining, in comments added to their surveys, that they do plenty more housework on the weekends.

"What a lot of women say is 'I'm carrying most of the load.' Even if husbands help, wives are the ones who are responsible for making sure everything is done," says Elizabeth M. Ozer, Ph.D., a psychologist and assistant professor in the Division of Adolescent Medicine at the University of California, San Francisco. "The tasks of paying household bills, managing kids' schedules, planning activities, grocery shopping and all those small details fall to me," laments one mother of three from Pennsylvania.

Lack of money is another significant stressor. When asked to choose from an assortment of causes, 38 percent of women cite financial worries as their greatest source of personal stress. This is significantly more than chose unreliable child care, relationship problems or other responses. Experts explain that lack of money breeds a sense of powerlessness in a person—a feeling that always increases stress. Not surprisingly, our survey found that the lower a woman's income, the more likely she is to suffer depression, to have left a job because of stress and to withdraw in a stressful situation. A woman with a low income is also more likely to work in a support-staff position, which gives her less control in her work life than other types of employment.

Lack of time is the final cause of stress because it eats into the very heart of what women hold dear. What they want is to spend their free hours enjoying their family. Yet many women feel they have too little time to spend relaxing with their husband and children because of work—and work stress—they bring home.

The Consequences of Stress

The chief result of job stress is the tension it causes at home. When asked to pinpoint specific ways work impacts home life, 34 percent of women say they must stay late and 28 percent say they bring work home.

But piles of papers aren't the only thing women bring home with them; they're carrying the accumulated tensions of the day. "Work tension is far more likely to spill over into the home than the other way around," says Dana Friedman of the Families and Work Institute. "That's because it's safer to express your frustration in the haven of your family, rather than risk annoying your boss by venting your stress at the office."

Husbands, especially, take the heat. Fifty-seven percent of working moms say they most frequently take their stress out on their spouses, and 46 percent say that a strained relationship with their partner is the biggest intangible toll of stress. "The stress has to go somewhere. And your spouse is the one who's always there," says psychology professor Ozer.

These strains also dampen working moms' pleasure in other areas. For instance, 35 percent of women cite diminished sex drive as one of the ways they react to stress. There are other physical and psychological costs as well, as the following charts illustrate:

How do you believe your stress shows up physically? Check up to three:	
Fatigue	65%
Headaches	46%
Muscle tension	45%
Overeating	32%
Insomnia	30%
Digestive problems	20%
Skin outbreaks	14%
Smoking	8%
Drinking	7%
High blood pressure	7%

How do you react psychologically to stress? Check up to three:	
Irritability	75%
Anger	42%
Depression	40%
Loss of energy	38%
Diminished sex drive	35%
Nervousness	20%

What Women Want

Here's the bind: Either women have money and no time to spend it, or they have time and no money to enjoy it. When asked what they think would most reduce their stress, over 60 percent of women specify either money or time, with 62 percent choosing money and 61 percent time for themselves. Whichever answer they select, women feel strongly that it would ease their stress. The director of a career center at a small university in Michigan writes, "I am one of the lucky ones. Money is not a worry. However, time *is* a worry. There is never enough time to finish projects at home or at work." On the other hand, a mother of two who works at a school in North Carolina says: "I love my family and I love being with them. But I wish money didn't rule our lives the way it does."

Much as women turn to their families for comfort and pleasure, however, many have learned to seek relief from stress beyond the boundaries of their intimate relationships. Our survey found that working mothers often look for a neutral ear when they need to share their troubles. In fact, 59 percent of women say they go to friends when they're feeling stressed, compared with 50 percent who go to their spouse and 32 percent who turn to a family member. This echoes other research on the buffering effect of what researchers call social support.

Next to spending time with family and friends, the most popular stress-relievers include sleeping (45 percent), watching television (41 percent), exercising (35 percent), shopping (31 percent) and eating (26 percent). Three out of five feel that the ways they relieve tension are effective.

On the work front, women see flexibility as the crucial stress-buster. Asked what they'd do if they could be the CEO of their company, almost half (48 percent) say they would first offer flextime for their employees. That's compared with 28 percent who say they'd first raise salaries. Women's desire for flexibility also showed up when we asked what they'd do if they won the lottery. More than a quarter (27 percent) said they would start their own business, an arrangement that offers a high degree of control and flexibility.

Learning to Speak Up

This much is clear: Working mothers are tough, and their take-charge attitude shows it. When faced with a stressful situation, 60 percent take action: They organize themselves a little better (32 percent) or work a little harder (28 percent). Almost nine out of 10 respondents (89 percent) feel in control of their work despite its pressures. This sense of being in charge of a situation is considered a hallmark of good mental health.

Women are not waiting for a revolution to make things better; they're coming up with their own ways to head off tension. For instance, Debbie Rambo of New Jersey, a mother of three who designs forms at a printing company, took a night shift so she could help her children with after-school activities and still have some time for herself.

Working mothers are also committed to the lives they've chosen. Stretched as they are by multiple demands, they'd rather have too much on their plates than too little. Though 79 percent of our respondents consider themselves more stressed than their mothers were, 69 percent say they are happier. They're glad to feel stress instead of depression, which plagued mothers of the fifties and sixties, says Rosalind C. Barnett, Ph.D., senior scholar at the Murray Research Center at Radcliffe College and coauthor of *She Works/He Works* (HarperCollins). In fact, our survey shows that women are far more likely to feel irritable (75 percent) than depressed (40 percent). "All evidence is that employed women, regardless of the work they do, experience higher levels of psychological well-being and lower levels of stress than women who stay home," Barnett notes.

These moms love the multifaceted lives they have created for themselves, but the price they pay is high. Must it be that way? One of the most telling discoveries of our survey is that only 11 percent of women say they ask for help when dealing with a stressful situation. Maybe it's time that changed.

Working moms know what kind of help they need. The next step is to ask for it—both at home and on the job. This may be hard to do, because they've worked so long to prove they can do it all—and all on their own. But they've made their point. Now it's time to share the struggle.

Reflections

1. What are your own physical and psychological responses to stress?

2. Have you experienced problems in your relationships as a result of stress? What were they? Were you able to resolve them?

3. Do you think that sources of stress are different for women and men? If so, in what ways?

4. In your experience do employed women "experience higher levels of psychological well-being and lower levels of stress than women who stay home"? What are the reasons?

According to the author, people should be allowed to spend their last days among their family and friends. She says, "Hospitals are great places to have surgery [and] to have babies. But not to die." What do you think?

I Want to Die at Home

By Anne Ricks Sumers

Last week my brother Rick died at home. I am so proud that we helped him die at home. Everyone should have as beautiful a death. Don't get me wrong—it is tragic and wrong that he should be dead. Rick was only 40, a nice, funny guy, a good husband, a dedicated lawyer, a father of a little boy who needs him. He didn't "choose death": he wanted desperately to live, but a brain tumor was killing him and the doctors couldn't do a thing. He had only one choice: die in the hospital or die at home.

Three months ago, doctors discovered cancer in both of Rick's kidneys. The kidneys were removed and he and his wife coped well with home dialysis. He returned to work, took care of his son, remodeled his kitchen and went to a Redskins game.

Then disaster struck. It all happened so quickly. On a Tuesday his wife noticed Rick was confused. By Wednesday he was hallucinating and an MRI showed that his brain was studded with inoperable cancer. Doctors said chemotherapy or radiation might buy him a few more weeks of life, but they couldn't predict whether he would become more lucid. When Rick was coherent, his only wish was to go home.

"Where are you, Rick?" said the intern, testing Rick's mental status.

"I don't know," he answered politely, "but I want to go home."

Rick was confused and disoriented, but he was fully aware that he was confused and disoriented. I showed him pictures of my children I keep on my key chain; he shook my keys and gently put then back in my hand: "You better drive. I'm too f___ed up to drive."

My brother was a strong guy. He kept getting up out of bed. The hospital staff tied him down. He was furious, humiliated, embarrassed, enraged, confused and frightened.

A hospital is no place to die. It's noisy and busy and impersonal; there's no privacy anywhere for conversation or a last marital snuggle; there's no place for the family to wait and nap; only one or two people can be with the dying person. The rest of the family hovers unconnected in a waiting room, drinking bad coffee, making long phone calls on pay phones in lobbies. It was exhausting.

Please don't misunderstand me; I like hospitals. Hospitals are great places to live, to struggle for life, to undergo treatments, to have surgery, to have babies. But not to die. If it's at all possible, people should die in familiar surroundings in their own beds.

Rick begged to go home. The doctors could offer us nothing more. Hospice care wasn't available on such short notice. We all knew it would be an enormous responsibility to take him home.

But Rick's wife is the most determined person I know. He wanted to go home and she wanted to take him home. Rick's doctor was empathetic and efficient; in less than two hours after we came to this decision, we were on the road in an ambulance, heading to his house.

Rick's last evening was wild, fun, tragic and exhilarating. Rick walked from room to room in his house, savoring a glass of red wine, eating a cookie, talking with his best friend, our mother and dad, our sisters and brothers. Neighbors stopped in with food and stayed for the conversation. Friends from the Quaker Meeting House stopped by. Cousins arrived.

It was like a Thanksgiving—good food, lots of conversation, but the guest of honor would be dead in a few days, or hours.

Although Rick was confused he wasn't frightened. Rick knew he was in his home, surrounded by friends and family. He was thrilled to be there. He ate. He cleaned. He was busy all evening, reminiscing, telling fragments of stories, neatening up, washing dishes, giving advice and eating well.

At the end of the evening he brushed his teeth, washed his face, lay down in his wife's arms in his own bed and kissed her goodnight. By morning he was in a deep coma.

All that long Saturday my family was together and we grieved. We watched over Rick. My father planted bulbs, daffodils and tulips, to make the spring beautiful for his grandson. My mother washed Rick's hair. My brothers and sisters and in-laws painted the porch banisters. Family, friends and neighbors came by to see him sleeping in his bed. Sometimes as many as 10 people were in his bedroom, talking, crying, laughing or telling stories about him, or just being with him—other times it was just his wife.

He took his last breath with his wife and his best friend beside him, his family singing old folk songs in the living room. He was peaceful, quiet, never frightened or restrained. Rick died far too young. But everyone should hope to die like this; not just with dignity, but with fun and love, with old friends and family.

Over the past week I have told friends about my brother's death. A few friends shared with me their regret about their parents dying prolonged and painful deaths alone in the hospital, sedated or agitated, not recognizing their children. Other people express fear of a dead body in the house—"Wasn't it ugly?" they ask. No, it looked like he was sleeping.

Rick's death was as gentle as a death can be. It worked because all parties—doctors, family, Rick and Rick's wife—were able to face facts and act on them. His doctors had the sense to recognize that no more could be done in the way of treatment and had the honesty to tell us. Rick's wife was determined to do right by him, whatever the burden of responsibility she was to bear. Our families were supportive of all her decisions and as loving and helpful as we could be.

Doctors must learn to let go—if there's nothing more to offer the patient then nothing more should be done—let patients go home. We all should plan for this among ourselves—preparing our next of kin. Families, husbands, wives need not be fearful. If a family member is dying in a hospital and wants to return home, try to find the means to do it.

And to all of the nurses, doctors and social workers in hospices: continue to do your good work. You have the right idea.

Reflections

1. What are your experiences with death and dying? Did they take place at home or in the hospital?

2. How would you feel about caring for a loved one who was dying at home? What resources might be available to you?

3. Where do you want to die? Why?

4. Did this article influence your thinking about where the dying process should happen?

According to family systems theory, all families are governed by rules. The rules of alcoholic families, however, are dysfunctional; they serve to maintain the status quo of alcoholism. In this excerpt from her book Another Chance, *Sharon Wegscheider discusses how the alcoholic wields power in the family and the unwritten rules that the family abides by.*

Rules in Alcoholic Families

Excerpted from *Another Chance*
By Sharon Wegscheider

Let us now look at what goes on in an alcoholic family in the light of...system principles. ...As you will recall, we described a system as "(1) made up of component parts that are (2) linked together in a particular way (3) to accomplish a common purpose."

We have only to consider the very first of those elements, the components, to realize that the alcoholic family is in for trouble. One member is hooked on an outside force and cannot move freely to maintain the system's balance. Because of that fact, he sends first ripples, then tidal waves of disturbance through the family system. Other members must adjust to this situation, and in time they become so damaged by the pressure and the postures they must assume to withstand it that they, too, start sending out waves of disturbance.

From this perspective, the alcoholic appears to be a very powerful person. And yet, [is he not] trapped and helpless, tossed on the ever stormier sea of his addiction? Here lies the paradox: *as the alcoholic gradually loses power over his own life and behavior, he wields more and more power over those of the people close to him.* Though he is increasingly dependent on them for support—emotional, social, and financial—he plays the dictator to get it. He controls what they say, what they do, what they think, even what they feel. The control is so constant, all-pervasive, and often subtle that they may not even be aware of it.

...the person who holds the power makes the rules. He literally designs the system, and he designs it in his own image. The problem is that in a dysfunctional family the most powerful person in terms of rule-making is also the most dysfunctional. In an alcoholic family that person is, of course, the Dependent. It should not surprise us, then, that the family takes on a group identity which mirrors that of the alcoholic, and soon everyone is displaying the psychological symptoms of his disease.

Predictably, alcoholic families are governed by rules that are inhuman, rigid, and designed to keep the system closed—unhealthy rules. They grow out of the alcoholic's personal goals, which are to maintain his access to alcohol, avoid pain, protect his defenses, and finally deny that any of these goals exist. Here a few of the rules that I encounter again and again in my work with these families.

Rule: The Dependent's use of alcohol is the most important thing in the family's life. For example, he is obsessed with maintaining his supply, and the rest of the family is just as obsessed with cutting it off. While he hides bottles, they search for them. While he stockpiles, they pour liquor down the drain. Like two football teams, their goals lie in opposite directions, but they are all playing the same game. They all plan their days around the Dependent's drinking hours—he to be sure that nothing interferes, they to frustrate his plans, or to arrange to be home in order to meet his demands, or to arrange not to be home in order to avoid his fury or possible embarrassment in front of their friends. The Dependent's use of alcohol is the overriding family concern around which everything else revolves.

Rule: Someone or something else caused the alcoholic's dependency; he is not responsible. Here the Dependent's increasing tendency to project his guilt and to blame someone else for his situation gets crystallized into a rule and imposed on the rest of the family. The scapegoat may be his wife or a child in trouble or a job he does not like—anything. Curiously, the scapegoat often goes along with the delusion and is overwhelmed with guilt and feelings of worthlessness.

Rule: The status quo must be maintained at all cost. It is easier to understand the extremely rigid way an alcoholic family responds to change by [thinking of the family as a] mobile. If the largest [piece] were to become snagged on some outside object, the string with which it is attached would pull taut and the supporting sticks would become rigid. Something similar happens when one family member gets snagged on a chemical. What's more, he is afraid to get unsnagged, for he feels that without it he cannot survive. So as rule-maker he makes sure that the sticks and strings of the family system stay rigid enough to protect him from change.

Rule: Everyone in the family must be an "enabler." When you ask members of an alcoholic family how they feel about the Dependent's drinking, they are of course quick to say that they would do anything to get him to stop. But all the while they are unconsciously helping him to continue—"enabling" him .... One person in the family plays the role of chief Enabler, but according to this unwritten rule, everyone else must do his part, too, to protect the Dependent and his dependency. They alibi for him, cover up, take over his responsibilities, accept his rules and quirks docilely rather than rock the boat. These actions may be defended on the grounds of love or loyalty or family honor, but their effect is to preserve the status quo.

Rule: No one may discuss what is really going on in the family, either with one another or with outsiders. This is exactly the sort of rule we would expect in a system as unhealthy and closed as an alcoholic family. Feeling threatened, the rule-maker tries to avoid, first, letting people outside know about family affairs—specifically, the degree of his dependency and the magnitude of its impact on his wife and children—and second, letting family members have access to new information and advice from outside that might undermine their willingness to enable.

Rule: No one may say what he is really feeling. This is a standard rule in severely dysfunctional families. The rule-maker is in so much emotional pain himself that he simply cannot handle the painful feelings of his family, which make his own even sharper. So he requires that everyone's true feelings be hidden. As a result, communication among family members is severely hampered. What there is tends to be rigid, distorted, and incomplete, the messages bearing little resemblance to the real facts and feelings that exist.

Eventually as his disease advances, the alcoholic completely represses his own feelings and ... unconsciously puts in their place false emotions that are less painful. These are the feelings that seem on the surface to prompt his actions. But to those who know him well, his performance is not quite convincing. They may respond as though they took his behavior at face value, but at some level they sense a second, subliminal message coming from the real self that he has repressed.

Dependent's True Feeling	Dependent's Behavior	Family Members' Feelings
Guilt, self-hatred	Self-righteousness, blaming	Guilt, self-hatred
Fear	Aggressiveness, anger	Fear
Helplessness	Controlling (of others)	Helplessness
Hurt	Abusiveness	Hurt
Loneliness, rejection	Rejecting	Loneliness, rejection
Low self-worth	Grandiosity, criticalness	Low self-worth

They are thus confronted with contradictory messages coming from different parts of the Dependent. One they hear with their rational minds, the other with intuition. They feel confused because the two messages are saying such totally different things:

"If these kids would show a little responsibility about money, I wouldn't have to be so hard on them." (I'm worried that I'm going to lose my job because I've called in sick so many Monday mornings.)

"If you were more affectionate, I wouldn't stay out late at night." (I know I'm not satisfying you—I don't know what's happened to me lately.)

"Why should I go to church? That new minister is only interested in money." (I'm no damned good. I can't face the minister, or the congregation either.)

Most often the false emotion expressed in his behavior is the opposite of the true emotion that lies underneath. Aggressiveness masks fear; blaming masks guilt; control masks helplessness. But, ironically, his behavior evokes the same painful feelings in family members that the Dependent is feeling underneath. In the table...we can see the dynamics of contagion by which family members gradually come to manifest the psychological symptoms of alcoholism.

Reflections

1. According to Wegscheider, how does the alcoholic gain control of the family?

2. What is the family's role in maintaining alcoholism?

3. What rules does the author discuss? Are any of these rules relevant to your own experience?

4. Are there other rules you can think of that might keep a family from recognizing and dealing with its problems?

5. What is your own experience with alcohol and alcoholism?

30

Although the story of Dorothy Rapp may seem particularly chilling as an account of subjugation and terror, it is in some sense a "classic" case in which many elements of the battered woman syndrome may be recognized.

The Deadly Rage of a Battered Wife

By Janet Bukovinsky

Bill Rapp spent a lot of time in the toolshed. He was meticulous about rearranging the tools hanging on the walls and about sharpening his log-splitting wedges and the chain saws he used to cut firewood. He worked as a welder in a Patterson, N.J., foundry.

Back in 1953, he'd moved his wife and baby from Colorado, where he was stationed as a staff sergeant in the Air Force, to New Jersey. They bought a red house in a nice residential neighborhood in Fair Lawn.

On March 26, 1981, Bill and Dorothy Rapp were sitting in their living room in front of the fireplace. Dorothy, a soft-faced woman whose mouth seemed to melt away at its corners, had a cold. She was bundled up in a flannel nightgown and her yellow terrycloth robe. The night was chilly, and their house had no central heating system. The Rapps burned wood to keep warm.

Bill was an alcoholic. He had already put away most of his nightly quart of vodka after dinner. He started picking on Dorothy that night, as usual, berating her because she hadn't brought in another load of wood chips for the fire.

He was drunk. There were plenty of wood chips in the house. Dorothy knew what was coming.

He threw her a punch that landed, hard, in her face, and he started pummeling her. Dorothy steeled herself to the familiar rhythms of his blows, yelling at him to stop, fighting back half-heartedly. When he did, she went into their bedroom and lay down in the dark on their double bed.

She was 48. Her life had been this way for 30 years, since she married Bill in Aurora, Col., when she was 18 and he 19. They had rented a little trailer there, and Dorothy soon became pregnant.

Bill's reaction to the news was an awful omen of things to come. It was the first time he beat her. He picked her up and threw her 36 feet—the length of the trailer. She slid on her side into the bedroom.

Bill Rapp's fury as he hurled his 110-pound wife across the trailer shocked her, but it was a kind of excruciating homecoming for Dorothy. At that point, Bill hadn't done anything to his wife that her father hadn't done to her mother.

"That's the way men are," counseled her mother the next day. "You married him for better or worse, until death do you part."

Dorothy returned to the trailer, and her life became one long bout of flinching and screaming.

He followed her into the bedroom, turned on the light and proceeded to drag her from the bed. "You didn't bring in enough wood chips," he screamed. "Go get more." Dorothy didn't argue.

When he threatened her life, as he had with a machete two weeks earlier, she would try to talk him out of his rage.

But tonight she sensed that the path of least resistance was the safest. She put on her slippers and went out into the darkness to fetch more wood chips. Then she went upstairs, where her 25-year-old son, James, was watching television.

He had heard the screams. They were nothing new to him. When the screams grew so agonized that James knew his mother needed help, he'd intervene, grappling with his father.

Tonight, he grimly advised Dorothy to stay out of Bill's way. The two of them heard the back door slam. Bill must have gone outside to his toolshed. She went downstairs, lay back down in their bed and dozed.

© 1982 by Janet Bukovinsky/*New Jersey Monthly*.

Bill was rarely sober now, except at work, and when he was drunk he seemed to prefer beating her to almost any other activity, including sex.

Dorothy spent most of her time in the house. She knew how to drive, but Bill took the car to work. She'd walk to the store or watch soap operas and crochet afghans, which she sold.

On that day in March, nearly one year ago, Dorothy awoke for the second time that evening. Bill had come back and was pounding on the back of her head—a favorite spot for abusive men's fists, since hair covers the bruises. He was punching her stomach and back, breaking several of her ribs, she was later to learn.

Dorothy curled her body into a ball, trying to make herself as small as possible. Bill grabbed a handful of her hair and tore it out. His inexplicable wrath was worse than usual.

When he came at her with the machete two weeks earlier, she'd run in terror to a neighbor's house to call the police, who had responded innumerable times to domestic argument calls at the Rapp residence. Bill met them out front, subdued and charming. Just a squabble, he told them.

Dorothy was standing on the sidewalk sobbing. "He's going to kill me!" she screamed. "Can't you just take him down [to the station] and talk to him?" The police advised her to leave and did so themselves.

When Bill ceased his latest round of beating, Dorothy lay, still curled up, with her back to him. She could see him in the mirror.

He took his hunting rifle from under the bed. It was a .30–.30 lever-action Winchester, a relatively lightweight gun and a legal hunting weapon. Dorothy heard a click as Bill pulled down the lever and inserted one bullet. He rested the gun against the wall near the top of the bed, on "his" side. "Your name is on that bullet," he used to tell her when he threatened her with the gun. "Don't you move or breathe until I come back," he seethed. "You're really going to get it." Then he went outside again.

Dorothy soon got out of bed. She picked up the rifle. She walked outside to the small porch. Bill was on his way back to the house. Dorothy couldn't see him well because a shelf on the porch obstructed her view. She stuck the barrel of the rifle through a space in the porch railing, aimed "in front of the sound of his voice" and touched the trigger.

The bullet with Dorothy Rapp's name on it struck Bill Rapp as he was walking toward his house with a chain-saw sharpener in his hand. It entered the front of his body, traveled on a downward path and exited under his left arm. His heart exploded.

Still holding the Winchester, Dorothy went into their bedroom and called the police. "Send somebody, an ambulance. I just shot my husband."

James came downstairs. "What was that?" he asked.

"I just shot Daddy."

At the trial of Dorothy Rapp last November, the prosecution charged that she had committed premeditated manslaughter—that, as a battered wife, she had reason enough to hate Bill and want to blow him away. Her attorney, Frank Lucianna, characterized her as the archetypal battered wife, paralyzed by fear and guilt ....

No socioeconomic group is immune. Connie Francis was a battered wife; so was Doris Day. The 1974 Nobel Peace Prize winner, the late Eisaku Sato, was publicly accused by his wife of abusing her prior to his nomination for the award.

The FBI believes that only one in 10 instances of marital abuse is reported. When the police respond to domestic violence calls, they simply aren't much help.

"Many police officers do not treat assault by a man upon his wife or female companion as a criminal act requiring arrest," said Clyde Allen, chairperson of the N.J. Advisory Committee to the U.S. Commission on Civil Rights.

Dorothy Rapp had found no ally in the police. "There were many times that the police discouraged me," she said. "I wanted to press charges, but they said I shouldn't do that, that I'd be wasting the court's time."

The police never saw Bill Rapp beating his wife. In New Jersey, acts classified as offenses, such as disorderly person charges (most applicable to wife-beating cases) must be witnessed by an officer for an arrest to be made.

At Dorothy Rapp's trial, Frank Lucianna suggested to the jury that the police had been such frequent visitors to the Rapp residence that they had simply stopped filing reports. He was determined to prove that Dorothy had acted in self-defense—that she had fired the shot to prevent Bill from killing her that night.

The fact that Dorothy was so completely subjugated by her husband was also important to her defense, in proving that she didn't have the strength to leave him. Julie Blackman Doron, a professor of psychology at Barnard College and an expert on battered wives, testified. She described Dorothy as a "psychologically cornered" woman, so entrenched in the morass of her marriage that she didn't know how to begin escaping her husband. Bill discouraged her from establishing contact with the outside world. She hadn't worked for years, and had no close friends. She was, said Doron, a prisoner in her own home.

The man who beats his wife is often, according to Dr. Arnold Hutschnecker, "a frail boy who did not give his ambition attainable goals in a world of reality. He had no conscious awareness of what it meant to feel secure, to be a man with self-confidence and self-esteem." Hutschnecker, Richard Nixon's former physician, was describing Lee Harvey Oswald, a chronic wife-beater.

The advice that Dorothy received from her mother in 1952, when Bill first beat her, was cited by Lucianna as part of the syndrome that kept her bound to Bill. Her mother viewed divorce as a failure of the most "womanly" role—keeping one's family together.

Had Dorothy been able to look beyond her mother's limited perspective, she might have been able to divorce Bill, though it's difficult to say whether a judge's decree and a piece of paper would have been enough of a deterrent when the urge to beat her washed over him.

Dorothy Rapp was found innocent of manslaughter on November 18. It was the first time that the battered-wife defense was successfully used in a murder trial in New Jersey. The jury of seven men and five women deliberated for less than two hours.

Reflections

1. What elements of the battered woman syndrome do you see in the story of Dorothy Rapp?

2. Do you think arrest or the threat of arrest would have kept Bill Rapp from battering his wife? Does it deter other violent men from harming (or killing) their partners?

3. If you knew someone in Dorothy Rapp's situation, what would you advise her to do?

4. How do you feel about the jury's verdict in the case?

31

As you read the following factual account of incest, note to what extent the family members resemble the corresponding profiles that you have encountered in your reading.

Breaking the Silence

By Meredith Maran

Jim's shoulders slump as his oldest daughter speaks. "My dad has been molesting me for about 10 years," Rachel begins calmly. "I knew it was wrong because he always said it was our little secret, that he'd go to jail if I told anyone. I loved my dad more than I loved anyone, and I kept thinking about how much my little sisters needed him—but I was afraid that he'd start doing it to them. So I made a deal with him: I wouldn't tell if he promised not to touch my sisters. He swore I was the only one, so I learned to cope. I learned to separate myself from my body—it was just my body lying there on that bed; my dad would never do those things to the real me. At first, I'd be upset for weeks every time he molested me. Near the end I'd gotten to where he'd finish and I'd get up and forget about it within an hour or two."

Sandy's [Rachel's mother] eyes are shut. Rachel glares again at Jim and continues.

"Then one day I heard Lisa crying in her room. I felt like my heart was on fire—I just knew what was wrong. She didn't want to say anything because he'd told her she was the only one. Finally she admitted that Dad had been handling her. When Sarah got home, Lisa and I talked to her. That's when we found out that he'd been molesting all three of us since we were babies.

"Mom wanted to call the police right away. But we told her we'd deny everything if she did. We didn't want to lose our dad, we didn't want our family to fall apart. We just wanted him to stop molesting us."

When Sandy confronted him, Jim cried and swore that it would never happen again, confiding for the first time that he'd been molested by his uncle for several years early in his childhood. Succumbing to the pleas of the man she'd loved for 20 years and the daughters she lived her life for, Sandy agreed to keep Jim's crime a secret. The next day, she installed deadbolts on each girl's bedroom door and ordered them to lock themselves in at night. And she swore to her husband that she'd have him arrested if he ever again attempted to have sex with their daughters.

"That next year was terrible for all of us," Sandy remembers. "I was losing my love for Jim, and I was terrified that he'd molest again. I kept asking the girls over and over if he was trying anything, and they kept promising me that everything was fine. I wanted to believe them so I shut out the little voice inside that told me something was wrong. I'd loved that man since I was 15 years old. I'd struggled so long to build the kind of family I thought we had—and I couldn't stand the thought of breaking it up, let alone telling the world that my husband was a child molester!

"Besides, Jim was our breadwinner. My income barely covered the groceries. So I prayed and prayed that somehow things would work out."

Sandy Rand's desperate decision not to report her husband's crime is one many wives of incest perpetrators make. Fear of devastating legal consequences, public humiliation and loss of income has made incest a painfully guarded secret for untold numbers of victims. But keeping sexual abuse a family secret only allows the abuse to continue. Without treatment, both the perpetrators and the victims may pass their behavior from generation to generation.

Sandy learned this lesson the hard way: Exactly one year after she'd agreed not to report the incest in her family, 12-year-old Sarah came to her in tears. Jim had been begging her for the

© 1984 by Meredith Maran, excerpted from West Magazine, May 20, 1984.

"special backrubs" that his older daughters now refused to give him.

"When Sandy confronted me the first time," Jim says now, "I knew I was in trouble, that I'd have to convince her to keep it in the family. But when she came to me about Sarah, I knew that I'd go to jail, that I'd never be with my wife and daughters again. Mostly I felt...relief. I'd been trying for so long to get Sandy to kick me out—seeing other women, picking fights—because I knew I'd never be able to stop as long as I lived with my daughters. As I watched her dial that phone, all that I could think was, "It's finally over."

The first call Sandy made was to Parents United. The counselor she spoke to explained that if Sandy gave Jim's name, PU would be legally obligated to report him to the police. Things would go better, the counselor said, if Jim turned himself in with the help of a lawyer who worked with PU on such cases. Rachel, Lisa, and Sarah didn't want to go to the meeting with the lawyer, but Sandy insisted. "Even after everything that had happened to them, the girls still didn't want Jim to go to jail. I hoped the lawyer would convince them that we were doing the right thing."

It was in the lawyer's office that the history of Jim's sexual abuse of his daughters was told in its entirety for the first time. He admitted everything: How he'd begun fondling Rachel when she was 4; how he'd developed an elaborate system of household chores that always left on daughter available to him; how he'd recently begun begging Rachel to let him penetrate her. He told how frightened he'd become when Sarah had her first asthma attack while he was molesting her. Hearing his three-hour confession, Jim's victims began to feel the rage they'd been swallowing for so many years.

"Something snapped inside me," Rachel remembers. "Before that, there was no wrong to him. He was my best friend, my wonderful dad. I believed him when he said he molested me because I was so pretty and he loved me so much. But as I was listening to the whole story, I was filled with hate. This man we all worshipped had been hurting us horribly all our lives. We told Mom to go ahead and send him to jail. At that moment, I hoped he'd never get out." She glares at Jim, whose reddened eyes meet hers. "Sometimes I still think he should spend his life in there."

Because Jim pleaded guilty, there was no trial.... He was sentenced to one year in Elmwood County Jail. His confession, clean police record, and willingness to undergo therapy qualified him for the county Child Sexual Abuse Treatment Program. Jim was placed on work furlough on the condition that he attend weekly counseling sessions and donate weekend labor to Parents United.

"My first night at Elmwood," Jim says, "was when it really hit me. If any other man had done those things to my daughters, I would have taken him apart. But that monster was me." His voice quavers; he swallows hard. "I would have given my life to make up to my daughters for what I'd done." When he was offered an early release after six month, his daughters protested unanimously; they wanted their father to finish his sentence. Jim passed up the release and spent the full year in jail.

Jim's removal from their home provided no respite for his wife and daughters. The girls turned their anger on Sandy, accusing her of not protecting them, of not having sex with Jim often enough, of not loving and understanding them as their father had done. "They called me names I'd never heard from their mouths before," Sandy says, looking down at Sarah. "They wouldn't obey me, they wouldn't listen or talk to me. They refused to go to counseling. Because of the molest, they'd been made wards of the court and I knew if they kept up their acting out I would lose them forever. So I did what I had to do to save what was left of our family."

What Sandy did was write a letter to the judge asking him to order her daughters into individual and family therapy. "At first we fought it, 'cause we were forced to go," Lisa remembers. "But then we realized that we couldn't keep our feelings inside forever. Plus, with our dad gone, we needed help dealing with Mom. It was like a war at our house."

Rachel adds quickly, "We still don't get along that well with her. It'll never be like it was with Dad. But at least now Theresa's [the therapist] taught us to communicate."

"I still love my wife deeply," Jim says, his eyes on Sandy. "But I know I've lost her. And my daughters...last week when we had our first session with the whole family, they told me they still love me but they hate me, too. I don't blame them one bit. I took the love they were giving me truthfully, I turned it around and abused them as women instead of protecting them as daughters. I knew I had a problem, but I blamed it on everyone else: my uncle who molested me, Sandy, my boss.... Now I can only blame myself. None of this would have happened if I'd gotten help before it was too late."

"It is too late," Sandy says quickly. "I still love this man. But my girls come first with me, and my girls want this marriage to end. So 20 years is dissolved. I've lost more than my husband because of sexual abuse. I lost my father years ago when he abused Rachel. I lost all my friends—they were mostly people Jim and I worked with and I didn't want them to know what we were going through. I lost my job because I fell apart emotionally. The worst thing I lost was my self-esteem, as a wife, a mother, a woman.

"Right now the only thing that gives me hope is the changes I see in myself and in my daughters. We're all independent now. I just graduated from a nursing program, so I'll never rely financially on anyone again. Rachel's got herself a good job so she can take care of herself, and she's engaged to a fine young man. Lisa—she's always been so quiet—speaks up more now. Sarah's learning not to blame herself for things that have been done to her. In a way, my daughters and I have grown up together through this tragedy. I've learned along with them how I was raised to be a victim, like my mother was before me. I know my daughters won't raise *their* daughters that way! And we've all learned to say no—a word that as females we were taught never to say."

Rachel jerks her head angrily. "You make it sound so easy, Mom. What about the other things we learned?" She turns to me. "Like not to trust anyone, ever. Like growing up believing that you pay for love with sex. Like having to lie to everyone all your life. I'm *still* lying, trying to explain why I won't let Dad walk me down the aisle on my wedding day. And Lisa's still two grades behind because she was afraid to come home to do her homework. And Sarah's still got her asthma. As long as my sisters and I are suffering because of what he did, I'll *never* forgive him."

"It hurts deep down to face up to what I've done to you," Jim says. He is sobbing now. "I'm so sorry…"

Rachel didn't let him finish. "You should feel sorry—I don't accept your apology. You can sit there and talk about facing up to it. Well, we face up to it whether we want to or not! Do you know that Lisa just lost her babysitting job because the parents found out she was molested? Do you know that Sarah still wakes up screaming almost every night? Do you know that none of us has a single friend who knows the truth about us! You just make me feel guilty with all your damn apologies, and I don't want to hear them anymore."

Sarah beings to wheeze. Sandy sits her up; Lisa digs through Sarah's book bag and pulls out a bronchial inhaler. As Sarah's panting slows, Lisa stares straight at me for a long moment.

"You said you wanted to know our story," she says. "Well, now you see how it is for us. And I want you to write it all down. I want every kid who's being molested right now to know that they shouldn't keep it a secret, that they deserve help and they can get help. And I want every man out there to know what it does to children to be sexually abused."

Tears run down her cheeks. "If we can keep this from happening to just one girl or one boy, maybe all this pain will be worth it."

Reflections

1. In what ways do Jim, Sandy, and their daughters resemble the profiles of family members that you have observed or encountered in your reading?

2. What factors contributed to the silence maintained by each family member?

3. What is the therapist's role in helping the family to heal?

4. What changes must each person undergo in order to continue the healing process?

5. Why do you think people are so reluctant to disclose the existence of incest within their families?

32

This article explores the sensitive area where cultures meet but do not mesh. It touches on our strong feelings about children, religion, morality, law, sexuality, and violence. As you read, try to view the situation from the points of view of the various parties involved.

Marriage or Rape?

By Peter Annin and Kendall Hamilton

The Nov. 9 wedding in Lincoln, Neb., was strictly traditional. A joint ceremony, it was planned by a 39-year-old gulf-war refugee from Iraq, who had arranged to marry his two eldest daughters, aged 13 and 14, to two fellow Iraqi refugees, aged 28 and 34. A Muslim cleric, flown in from Ohio, performed the rites before a small gathering of relatives, and then, as custom dictates, the men and women celebrated the new unions separately, at adjacent houses. The trouble began a few days later, when the older daughter ran away. Concerned, the girl's father and husband contacted her school, which in turn called police. And that's when Islamic tradition ran up against Nebraska law. Not long after the police were done asking questions, the two girls were placed in foster homes, and five adults stood accused of crimes from child abuse to rape.

The controversy has rattled the people of Lincoln, an easygoing college town of 170,000, and revealed a cultural void as vast as the Middle Eastern desert. The police say that not only were the marriages consummated, violating a state law banning sex between anyone over 18 and anyone under 16 regardless of consent, but also that the girls were forced to have sex with their new husbands against their will—a charge the Iraqi men strongly deny. The case even temporarily usurped the hallowed Cornhuskers football team as the hottest topic in town. If some longtime Lincoln residents were taken aback by the arranged marriages in their midst, the Iraqis were equally puzzled by the legal fallout. Such marriages, even among the very young, are common in much of the Muslim world—and those in the city's close-knit Iraqi community say the girls entered the unions willingly. "I spoke to both girls in the days before the wedding," says Aeeda Al-Khafaji, a 30-year-old mother of seven who herself wed at 12 in Iraq. "They were both happy and excited. There was no problem at all." Still, the Iraqis say they're willing to respect the laws of their new land. "We are shocked by these allegations, and we are going to be much more careful [about the age at which girls marry]," says Mohamed Nassir, head of the Lincoln Islamic Foundation. "The problem is, we really didn't know what the law was."

But whose job is it to tell them? As refugees, the Iraqis were relocated by the federal government, which contracts with relief agencies like Catholic Social Services in Lincoln to coordinate housing, financial assistance, medical care and general orientation. The agency would release only a terse statement that "we expect all refugees assisted by our agency to obey all proper civil laws of our country." But Sanford J. Pollack, the attorney for the girls' father, says it's naïve to ask refugees—who face cultural and language barriers—to absorb legal rules on their own. "When we bring people to the United States," Pollack says, "we need to educate them about our laws and customs."

Now Lincoln's refugees are getting a crash course, courtesy of the city prosecutor's office. The husbands, Majed Al-Timimy, 28, and Latif Al-Hussani, 34, have been charged with rape and could face up to 50 years in prison. For his role in arranging the marriage, the girls' father has been charged with child abuse. Their mother is accused of contributing to the delinquency of a minor. Also caught up in the case is 20-year-old Mario Rojas, with whom the elder daughter was found living after her disappearance. He has been charged with statutory rape. All of the accused have entered pleas of not guilty. Lincoln's district attorney has declined to comment on the case, but one prosecutor, Jodi Nelson, has said that ignorance is no alibi. "You live in our country, you live by our laws," she says. Donations toward the defense have been received from as far away as Saudi Arabia, where the case has made front-page news.

To experts in Middle Eastern affairs, the situation is frustrating. "What you have here is cultural incomprehension," says Prof. Rashid Khalidi of the University of Chicago. "And the problem is that the law is not a very subtle instrument in dealing with these kinds of complex cultural issues." Khalidi calls the uproar in Lincoln "a prime example of a case that begs for a nonlegal resolution." Perhaps. But for the accused, the decision may rest ultimately with another bit of unfamiliar culture: a jury.

Reflections

1. What are the basic elements of the "cultural incomprehension" described in the article?

2. Do you feel there is a moral "right" and "wrong" in this situation? What factors is your reasoning based on?

3. Imagine what the following people in this case would have to say about their roles and motivations: the teenage girls, the husbands, the girls' mother, the girls' father.

4. If you were the judge, how would you resolve this case?

5. Is there a way to prevent this type of misunderstanding in the future?

Judith Wallerstein's work with the children of divorce has been extremely influential. In this ten-year clinical follow-up of children who were six to eight years old at the time of their parents' divorce, Wallerstein finds that as the children enter young adulthood, about half continue to be affected by the divorce.

Children of Divorce: Ten Years After

By Judith Wallerstein

Attitudes and Feelings toward Past and Present

In their unhappiness, their loneliness, their sense of neediness and deprivation, the youngsters now 16 to 18 years old suffered more than the other age groups in the study. The divorce was regarded as the central experience in their lives by over half of these young people, who spoke longingly of their lives in the intact, predivorce family. Some volunteered that their difficulties had escalated through the years, indicating that they had been more protected from parental quarrels when they were younger. An overwhelming majority, boys and girls alike, spoke wistfully of their longing for an ideal intact family. These feelings were unrelated to their judgment of the wisdom of their parents' decision to divorce. The majority, in fact, regarded their parents as incompatible, the divorce as irreversible, and the relationship between the parents as beyond repair. Nevertheless, they often explained a change in one parent's mood as a reflection of a major change in the other parent's life, as if it were self-evident that strong, invisible ties between the divorced parents had lasted over the years or as if their view of their parents, as a couple, had endured. Tom told us, "Mom's been bitchy since Dad's new wife had a baby."

Their sense of powerlessness in the face of the major event in their lives was striking. This was conveyed along with a sad, somewhat stoic acceptance of their experience. Alice said:

I don't know if divorce is ever a good thing, but if it is going to happen, it is going to happen. If one person wants out, he wants out. It can't be changed. I get depressed when I think about it. I get sad and angry when I think about what happened to me.

The recurring theme was loss of the father, even though there seemed to be almost no link between the father for whom they yearned and the actual father, to whom access was entirely open and who in many instances lived nearby. When asked if she had gained anything from the divorce, Olga said emphatically, "Absolutely no." Asked what she had lost, the girl responded, "My father. Being close to my dad. I wish it were different, but it is not going to change. It is too late." Olga's father lived nearby and she saw him monthly. He paid little attention to her at these meetings, preferring his son.

Larry's father also lived close by. The boy and his mother had asked the father to take Larry into his home, but the father had refused. Larry told us:

Life has been worse for me than for other kids because I was a divorced kid. Most of my friends had two parents and those kids got the things they wanted. Not having a dad is tough for me. I wanted to live with him but he would not take me. He never told me why.

Karl, who lived with his mother and stepfather and two older sisters, said, "I needed a father, not because I liked him more, but because there was no one in the home like me. That's my true feeling." We were interested that Karl did not include his stepfather in his perspective of someone who would be like him within the home.

Perhaps one clue to the distress expressed by so many youngsters was their sense of the unavailability of the working mother. It was not unusual for the youngster to equate the mother's unavailability with uninterest or rejection. Chuck said that his mother did not care for him. She was busy working all the time. "She does not pay any attention to me. I want her to be a mom with an interest in what I am doing with my life, not just a machine that shells out money." It would appear that the longing for the father may reflect not only feelings about the father, but also the feeling so many of these young people had of being rejected by a busy, working mother who was not available to them, and the overall sense that so many shared of not having been provided with the close support that they wished for and needed from the family during their childhood and adolescent years. It is possible that, in the same way that the youngest group of children at the ten-year mark were preoccupied with the idealized family of their fantasies, these youngsters were preoccupied with the lost father as the symbolic equation of the divorce. Their preoccupation with the father, unrelated to the actual quality of their father-child relationships, repeats their responses of ten years earlier.

Finally, although resentment was surely implicit in many of the statements of these young people, overt expression of anger toward parents, especially toward fathers, was uncommon. One of the distinguishing characteristics of this entire group, particularly the boys, was the muting of anger. This psychological stance parallels the limited acting out among these young people. Speaking of his father, who refused to support his college career unless the boy came to live with him and the new stepmother, even though the tension in the remarriage was very high, Andy said:

> He won't change. He won't allow anything else to come into his mind. I just learn to accept it. Sometimes I feel sorry for him. He really needs a wife. He and his new wife just don't get along.

Sometimes the need to avoid anger pushed youngsters into far-fetched apologies for their parents. Kelly told us how angry her mother had been when Kelly asked to live with her father. She had been ostracized by her mother, she was even asked to eat at a separate table following her request. Kelly confided, "I can't really blame my mom completely. She stayed up all night to take care of me when I was one year old."

Attitudes toward the Future

Most of these young people believe in romantic love. With few exceptions, they expect to fall in love, marry, and have children of their own. Like most of the other young people in this study, their values are conservative. They do not regard the divorced family as a new social norm. They consider divorce a solution that reflects marital failure, one that should be used only as a last resort when there are children. They agree on divorce when there is physical violence in the family. They believe, on balance, that divorce helps parents, not children. Their values include fidelity and life-long commitment. It is painful for them to acknowledge what many of them know, that one of their parents had been unfaithful during the marriage. Anger at a parent whom they know to have been promiscuous can be very bitter. Describing her father's many affairs, Betty said tartly,

> Some day he won't be able to cultivate all those 21-year-old girls and life will catch up with him. He is going to be an old man and he will be punished.

Close to two thirds of these young people were apprehensive about the possibility of disruption in their own future marriages. Girls were especially fearful that their marriages would not endure. A recurrent theme was a sense of vulnerability and fear of being hurt by romantic relationships. Talking about the future, Brenda said,

> It is hard to make a commitment. All the work and all the trust that is involved. I don't want to get married and do what my mom did. I don't know if marriage will last or not. How can you be sure that marriage will last? I hate to think of what will happen. I am afraid. I am afraid of being hurt. That's why I am a loner.

One half of the boys and girls were fearful of being betrayed, not only in their future, but in their present relationships as well. Of all the age groups, these youngsters were most worried about repeating their parents' relationship patterns and mistakes. Nancy said, "I always find myself attracted to guys who treat you bad." Maureen said, "A problem I have is not being able to show my feelings. I am afraid that they might get stepped on. Once in a relationship, I feel that I will be afraid of losing it if I get attached." Teresa said, "There is a lonely, shaky part of me. I am afraid of what happened to my parents happening to me."

A repeated fear among the boys, somewhat different from the fear of betrayal that the girls emphasized, was their fear of being unloved. Zachary said, "The divorce made me cautious of my relationships. Whenever I meet a girl, I have the unconscious feeling that when she gets to know me she will not love me."

A substantial minority of these young people, however, felt that they were but little influenced by their parents' failures and were relatively confident about the future. Several sought role models elsewhere in order to build their expectations on a solid ground. Barbara told us that she had selected her grandparents and their long-lasting marriage as the model for her future plans.

> I look at my grandparents who have been married 40 years. I don't look at divorce. It does not cross my mind. I don't think, if I got married, I would get a divorce. I don't worry about losing relationships.

Several young people told of their pleasure with the intact families of their boyfriends or girlfriends and their reassurance in finding examples of stable, happy marriages. Susan told us,

> My boyfriend has a family and I love his family. His parents have been married a long time. They are Irish. The kids all live at home. It is fun to be around them. I am always over there.

Independence

The forthcoming move toward independence created a great deal of anxiety in these young people. Although many . . . spoke proudly of their independence as a positive outcome of their parents' divorce, their behavior was often discrepant with their pronouncements. Mary told us: "Nobody helped me. Just my own determination and my friends." She had learned from a soccer accident that, "If you want to play, you play in pain." Diana said, "The outcome of the divorce was that to survive I had to be independent." Others told us how they had learned to solve problems on their own. Yet, although most of them were employed at least part-time and taking responsibility for themselves to a high degree, few spoke of wanting to establish themselves independently, and only three of these young people had left home to live on their own. Several of the youngsters who spoke bravely of their independence had suffered intensely during their freshman year at college and sought to return home.

A dream of Katherine's, which occurred during her freshman year at an out-of-town college, reveals some of the difficulties faced by these young people who were attempting to establish independence at a time when their own insecurities and need for parenting dominated their thoughts. Katherine told us,

> I had a dream at mid-term. In real life, I told a friend that the first thing I would do when I would get home was that I would hug my dad, and that would be proof that he loved me, and that I loved him. In the dream I came home and my dad wasn't there. The person who met me said, "Haven't you forgotten that your dad is dead?"

Katherine said that what disturbed her most was not that her dad was dead, but that there was no one to hug her, and she had worried so much about getting home for a hug.

It appears that independent behavior, and the pride young people feel in it, can mask an intense hunger for further nurturance and powerful feelings of not being sufficiently nurtured to make it on one's own without "playing in pain."

Reflections

1. What effects does Dr. Wallerstein observe in the children of divorced parents described in the excerpt?

2. Do you believe the feelings expressed by these young people are significantly different from those of individuals who parents have not divorced? Explain your reasoning.

3. What long-term effects of divorce have you observed in your own life (or the lives of people close to you), if any?

4. Can you think of ways in which divorcing parents might help their children to better deal with the effects of divorce?

In this excerpt from his book Life Without Father, *sociologist David Popenoe cites the absence of fathers as an important cause in "the collapse of children's well-being." Do you believe fathers are essential for the rearing of psychologically healthy children?*

When Dads Disappear

Excerpted from *Life Without Father*
By David Popenoe

The decline of fatherhood is one of the most basic, unexpected and extraordinary trends of our time. Its dimensions can be captured in a single statistic: In just three decades, between 1960 and 1990, the percentage of children living apart from their biological fathers more than doubled, from 17 percent to 36 percent. By the turn of the century, nearly 50 percent of American children may be going to sleep each evening without being able to say goodnight to their dads.

No one predicted this trend; few researchers or government agencies have monitored it; and it is not widely discussed, even today. But the decline of fatherhood is a major force behind many of the most disturbing problems that plague American society: crime, premature sexuality and out-of-wedlock births to teenagers; deteriorating educational achievement; depression, substance abuse and alienation among adolescents; and the growing number of women and children in poverty.

Even as the calamity unfolds, our cultural view of fatherhood itself is changing. Few people doubt the fundamental importance of mothers. But fathers? More and more, the question of whether fathers are really necessary is being raised. Many would answer no, or maybe not. And to the degree that fathers are still thought necessary, fatherhood is said by many to be merely a social role that others can play: mothers, partners, stepfathers, uncles and aunts, grandparents. Perhaps the script can even be rewritten and the role changed—or dropped.

There was a time in the past when fatherlessness was far more common than it is today, but death was to blame, not divorce, desertion and out-of-wedlock births. Almost all of today's fatherless children have fathers who are alive, well and perfectly capable of shouldering the responsibility of fatherhood. Who would ever have thought that so many men would choose to relinquish them?

Not so long ago, the change in the cause of fatherlessness was dismissed as irrelevant in many quarters, including among social scientists.

Children, it was said, are merely losing their parents in a different way than they used to. You don't hear that very much anymore.

A surprising finding of recent social-science research is that it is decidedly worse for a child to lose a father in the modern, voluntary way than through death. The children of divorced and never-married mothers are less successful in life by almost every measure than the children of widowed mothers. The replacement of death by divorce as the prime cause of fatherlessness, then, is a monumental setback in the history of childhood.

Until the 1960s, the falling death rate and the rising divorce rate neutralized each other. In 1900, the percentage of all American children living in single-parent families was 8.5 percent. By 1960, it had increased to 9.1 percent.

But the decline in the death rate slowed, and the divorce rate skyrocketed. "The scale of martial breakdowns in the West since 1960 has no historical precedent that I know of, and seems unique," says Lawrence Stone, the noted Princeton University family historian. "There has been nothing like it for the last 2,000 years, and probably longer."

Reprinted with the permission of The Free Press, a Division of Simon & Schuster from *Life Without Father: Compelling New Evidence that Fatherhood and Marriage are Indispensable for the Good of Children and Society* by David Popenoe. Copyright © 1996 by David Popenoe.

Impact on 'baby bust' kids

Consider what has happened to children. Most estimates are that only about 50 percent of the children born during the 1970-84 "baby bust" period will still live with their natural parents by age 17—a staggering drop from nearly 80 percent.

In theory, divorce need not mean disconnection. In reality, it often does.

One large survey in the late 1980s found that about one in five divorced fathers had not seen his children in the past year, and fewer than half of divorced fathers saw their children more than several times a year.

A 1981 survey of adolescents who were living apart from their fathers found that 52 percent had not seen them at all in more than a year; only 16 percent saw their fathers as often as once a week.

The picture grows worse. Just as divorce has overtaken death as the leading cause of fatherlessness, out-of-wedlock births are expected to surpass divorce later in the 1990s.

Across time and cultures, fathers have always been considered essential—and not just for their sperm. Marriage and the nuclear family—mother, father and children—are the most universal social institutions in existence. In no society has the birth of children out of wedlock been the cultural norm. to the contrary, a concern for the legitimacy of children is nearly universal.

At the same time, being a father is universally problematic for men. While mothers the world over bear and nurture their young with an intrinsic acknowledgment and, most commonly, acceptance of their roles, the process of taking on the role of father is often filled with conflict and doubt.

The source of this sex-role difference can be plainly stated. Men are not biologically as attuned to being committed fathers as women are to being committed mothers. The evolutionary logic is clear.

Women, who can bear only a limited number of children, have a great incentive to invest their energy in rearing children, while men, who can father many offspring, do not. Left culturally unregulated, men's sexual behavior can be promiscuous, their paternity casual, their commitment to families weak.

This is not to say that the role of father is foreign to male nature. Far from it. Evolutionary scientists tell us that the development of the fathering capacity and high paternal investments in offspring—features not common among our primate relatives—have been sources of enormous evolutionary advantage for human beings.

In recognition of the fatherhood problem, human cultures have used sanctions to bind men to their children, and of course the institution of marriage has been culture's chief vehicle.

Two parents are best

In my many years as a sociologist, I have found few other bodies of evidence that lean so much in one direction as this one: On the whole, two parents—a father and a mother—are better for a child than one parent.

There are, to be sure, many factors that complicate this simple proposition. We all know of a two-parent family that is truly dysfunctional—the proverbial family from hell. A child can certainly be raised to a fulfilling adulthood by one loving parent who is wholly devoted to the child's well-being. But such exceptions do not invalidate the rule.

The collapse of children's well-being in the United States has reached breathtaking proportions. Juvenile violent crime has increased sixfold, from 16,000 arrests in 1960 to 96,000 in 1992. Eating disorders and rates of depression have soared among adolescent girls.

Teen suicide has tripled. Substance abuse among teenagers, although it has leveled off in recent years, continues at a very high rate. Poverty has shifted from the elderly to the young.

One can think of many explanations for these unhappy developments: the growth of commercialism and consumerism, the influence of television and the mass media, the decline of religion, the widespread availability of guns and addictive drugs, and the decay of social order and neighborhood relationships.

None of these causes should be dismissed. But the evidence is now strong that the absence of fathers from the lives of children is one of the most important causes.

Economic ripples

The most tangible and immediate consequence of fatherlessness for children is the loss of economic resources. By the best recent estimates, the income of the household in which a child remains after a divorce instantly declines by about 21 percent per capita on average, while expenses tend to go up.

What do fathers do? Much of what they contribute to the growth of their children, of course, is simply the result of being a second adult in the home. Bringing up children is demanding, stressful and often exhausting. Two adults cannot only support and spell each other; they can offset each other's deficiencies and build on each other's strengths.

Recent research has given us much deeper—and more surprising—insights into the father's role in child rearing. It shows that in almost all of their interactions with children, fathers do things a little differently from mothers.

What fathers do—their special parenting style—is not only highly complementary to what mothers do, but is by all indications important in its own right for optimum child rearing.

For example, an often-overlooked dimensions of fathering is play. From their children's birth through adolescence, fathers tend to emphasize play more than caretaking.

The father's style of play seems to have unusual significance. It is likely to be both physically stimulating and exciting. With older children, it involves more physical games and teamwork requiring the competitive testing of physical and mental skills.

Mothers tend to spend more time playing with their children, but theirs is a different kind of play.

Moms stress security

Mothers' play tends to take place more at the child's level. Mothers provide the child with the opportunity to direct the play, to be in charge, to proceed at the child's own pace. At play and in other realms, fathers tend to stress competition, challenge, initiative, risk-taking and independence. Mothers, as caretakers, stress emotional security and personal safety.

Becoming a mature and competent adult involves the integration of two often-contradictory human desires: for communion, or the feeling of being included, connected, and related, and for agency, which entails independence, individuality and self-fulfillment. One without the other is a denuded and impaired humanity, an incomplete realization of human potential.

Just as cultural forms can be discarded, dismantled and declared obsolete, so can they be reinvented. In order to restore marriage and reinstate fathers in the lives of their children, we are somehow going to have to undo the cultural shift of the past few decades toward radical individualism.

Marriage must be re-established as a strong social institution. The father's role must also be redefined in a way that neglects neither historical models nor the unique attributes of modern societies, the new roles for women and the special qualities that men bring to child rearing.

Reflections

1. What are some of the effects of fatherlessness described by Popenoe?

2. Do you believe these effects are caused mainly by the absence of fathers, or could there be other reasons? Explain.

3. Do you believe it is possible for a single parent (either mother or father) to raise well-adjusted children? If so, under what conditions might this be possible?

4. Why do you suppose so many fathers choose to give up the responsibilities of fatherhood?

5. Based on your own experiences (in either a single-parent or two-parent family—or both), what do you believe the father's role should be? How can society encourage, support, or require fathers to be involved in rearing their children?

35

There is much to be said about retaining friendly, cooperative relationships between former spouses. As you read this article, be aware of the advantages and disadvantages and the major characteristics which divided or united couples.

Friends Through It All

By Elizabeth Stark

The common image most people have of divorced couples is warring partners fighting over finances and custody issues. This stereotype, perpetuated in jokes, movies and television, has been accepted by society as the norm. The prevailing attitude is that "there's something a little crazy if you still have a good relationship with your ex-spouse," says Eleanor Macklin, a psychologist at Syracuse University. Even the words we use to describe divorced spouses, "ex" and "former," lack the capacity to indicate any kind of surviving relationship, [Constance] Ahrons points out....

Many psychologists, such as Constance Ahrons and Macklin, now view divorce as more of a transition than an ending and are focusing on how to help families adapt. It's estimated that at least one of every three children growing up today will have a step-parent before they reach the age of 18. With so many parents getting divorced and remarrying other parents, the traditional concept of family is no longer adequate, Ahrons says. She uses the term "binuclear family" to refer to the families created by divorce and remarriage.

Many people mistakenly hope to "reconstitute" the nuclear family when they remarry and end up excluding the nonresidential parents, says Margaret Crosbie-Burnett, a psychologist at the University of Wisconsin-Madison. "In this culture we tend to think that a child has to choose one real mom or dad. Of course it doesn't work," she says. "Kids can easily accept two sets of parents." Crosbie-Burnett found in her study of 87 "stepfather" families—families in which the mother had remarried and had custody of the children from the first marriage—that those who maintained a friendly, or at least businesslike, relationship with their ex-spouses were much happier than those with hostile or unfriendly relationships.

Ahrons believes that it's perfectly healthy for divorced couples to have feelings of kinship and that they shouldn't be discouraged. She has been following 98 divorced couples for the past five years in her Binuclear Family Project, interviewing not only former wives and husbands but any new spouses or "spouse equivalents" as well at one year, three years and five years after divorce. All of the divorced couples had had children together, and the average duration of their marriages had been 10 years. To be included in the study both spouses had to live in Dane County, Wisconsin, and the noncustodial parent must have seen the children at least once in the past two months. Of the 98 pairs, 54 had maternal custody, 28 had joint custody and 16 had paternal or split custody—in which each parent has custody of different children.

Based on the frequency and quality of their interactions, couples were divided into...four groups.... Perfect Pals made up 12 percent of the sample. These couples enjoyed each other's company and tended to stay involved in each other's lives, phoning to share exciting news, for example. They were also very child-centered and tried to put their children's interests ahead of their own anger and frustration. Many had joint custody, and none were remarried or living with someone.

Ahrons interviewed one couple who actually shared a duplex apartment so that their children could come and go freely between their homes. The only drawback, they admitted, was lack of privacy, and they suspected that if one took on a new spouse or live-in lover, the arrangement might become awkward.

The largest group was Cooperative Colleagues, who made up 38 percent of the couples. They were not as involved in each other's lives as Perfect Pals, but they managed to minimize potential conflicts, to have a moderate amount of interaction and to be mutually supportive of each other. Ahrons sees them as the most realistic positive role model for divorcing couples.

Angry Associates, who accounted for 25 percent, also had a moderate amount of interaction, but the interactions were fraught with conflict. This group was unable to untangle spousal and parental issues, thus setting the scene for fighting when they dealt with each other.

The archetypal feuding partners, Fiery Foes, made up 24 percent of the couples. They had as little interaction as possible, argued when they did interact and did not cooperate at all in parenting. Ahrons suspects that this group may be underrepresented in her study since some of her criteria for inclusion would have ruled out many of the most antagonistic couples.

One of the major characteristics distinguishing the four groups was how they handled anger. Even Perfect Pals still harbored some anger over the divorce, but it was not a major part of the relationship and they were able to talk it out when it erupted. One said the relationship was "like having a good friend that you can still be angry with." Cooperative Colleagues and Angry Associates had about the same amount of anger, but Cooperative Colleagues could separate the old spousal issues from parental ones, while Angry Associates could not. Fiery Foes' anger was so overwhelming that it prevented any civil interaction.

Time, apparently, did not heal the wounds of divorce for the most antagonistic. Ahrons found that Angry Associates and Fiery Foes were just as angry about the divorce three years later as they were one year later. And all couples became more distant over time: interaction and positive feelings decreased over five years ....

It's clear that a cooperative and friendly relationship between ex-spouses is beneficial to their children, but is it harmful to later marriages or romances? According to a study by Macklin and Carolyn Weston, a psychology doctoral candidate at Syracuse University, friendship between ex-spouses it not necessarily bad and can even be beneficial to a second marriage.

Macklin and Weston interviewed members of 60 stepfather families; none of the stepfathers had any of their own children living in the household. Based on an earlier study by psychologist W. Glenn Clingempeel at Temple University, Macklin and Weston suspected that those with moderate contact would have the happiest marriages. But they found that the more contact a woman had with her former husband, the happier the new marriage, as long as both she and her new husband agreed about the nature and frequency of contact ....

Why did frequent contact make for happier second marriages? First of all, Macklin says, "It works both ways. A happier new marriage may allow more contact with the ex-spouse." But the main reason, she explains, is parenting. Regular contact between former spouses usually ensures that the father will be more financially supportive and involved in routine coparenting. And new husbands are usually happier when they do not end up shouldering the entire burden of raising their stepchildren.

Marion, one of the women Macklin and Weston interviewed, is a case in point. She was devastated when her husband left her for another woman. But she decided that she had to put it all behind her and two years later began living with David, to whom she is now married. Eight years after the divorce, she and her ex-husband, Jack, are now on very good terms. Jack comes over sometimes to watch television with his two teenage sons and David, goes to baseball games with them and even has keys to the house. Both Marion and David feel that it is very important for her sons to be actively involved with their father. They say that the boys get the both of both worlds, two fathers.

In Marion's case, continuing friendship with her former husband did not stop her from remarrying someone else. Are there situations in which lingering attachments prevent divorced men and women from replacing the former spouse? Ahrons found that Perfect Pals, those who interacted frequently, had never remarried. But she feels this was due to the couples' preoccupation with their children, not with each other. "These parents are so wrapped up in their children, they don't want to remarry for fear it will upset the balance." Another possibility, she says, is that outsiders see this tight-knit unit and are put off. For these couples, parenting issues are probably what draw them together....

Although some people question the point of continuing a relationship if a couple has no desire to reconcile, Gary Ganahl points out that "if a couple has children there is no question that they are going to have a continuing relationship." Keeping the relationship supportive and friendly benefits the children and in many cases can make life happier for the divorced couple.

On the other hand, Ahrons says, there are some situations where friendship between ex-spouses is impossible. The main point, according to all these researchers and specialists, is that it is not necessarily harmful for ex-spouses to be friendly. "Right now there are no positive norms or messages from society about divorce," Ahrons says. "Most people don't want to give up their total history, but there is very little sanction or support for their relationship to continue."

During the course of her interviews Ahrons discovered that "no one had asked most of these couples positive questions about their divorce before. They had never thought of the concept of a binuclear family." She would like to provide a positive role model for couples who want to remain friendly. She thinks if people reacted to divorce with the expectation that relationships don't necessarily have to end, "maybe more spouses would be encouraged to have amicable relationships."

The main thing for everyone—therapists, friends, family and couples—to keep in mind, Ganahl says, is that "relationships change, but rarely dissolve after divorce."

Reflections

1. Have you been able to maintain friendly relationships with former dating or marital partners?

2. What factors have created to making this possible or impossible?

3. When this was not possible, what got in the way?

4. Under what circumstances would you recommend a couple not remain friends?

In this excerpt from Carol Stack's classic ethnographic study of African-American ghetto poor, the author reveals the informal kinship system that evolves to help care for children of unmarried mothers. As you read through this article, pay attention to how your values and attitudes about child-rearing compare with those of the community described.

Single Mothers and Helping Kin in an African-American Community

Excerpted from *All Our Kin*
By Carol Stack

Billy, a young black woman in The Flats, was raised by her mother and her mother's "old man." She has three children of her own by different fathers. Billy says, "Most people kin to me are in this neighborhood, right here in The Flats, but I got people in the South, in Chicago, and in Ohio too. I couldn't tell most of their names and most of them aren't really kinfolk to me. Starting down the street from here, take my father, he ain't my daddy, he's no father to me. I ain't got but one daddy and that's Jason. The one who raised me. My kids' daddies, that's something else, all their daddies' people really take to them—they always doing things and making a fuss about them. We help each other out and that's what kinfolks are all about."

Throughout the world, individuals distinguish kin from non-kin. Moreover, kin terms are frequently extended to non-kin, and social relations among non-kin may be conducted within the idiom of kinship. Individuals acquire socially recognized kinship relations with others through a chain of socially recognized parent-child connections. The chain of parent-child connections is essential to the structuring of kin groups.

My experience in The Flats suggests that the folk system of parental rights and duties determines who is eligible to be a member of the personal kinship network of a newborn child. This system of rights and duties should not be confused with the official, written statutory law of the state. The local, folk system of rights and duties pertaining to parenthood are enforced only by sanctions within the community. Community members clearly operate within two different system: the folk system and the legal system of the courts and welfare offices.

Motherhood

Men and women in The Flats regard child-begetting and childbearing as a natural and highly desirable phenomenon. Lottie James was fifteen when she became pregnant. The baby's father, Herman, the socially recognized genitor, was a neighbor and father of two other children. Lottie talked with her mother during her second month of pregnancy. She said, "Herman went and told my mama I was pregnant. She was in the kitchen cooking. I told him not to tell nobody, I wanted to keep it a secret, but he told me times will tell. My mama said to me, 'I had you and you should have your child. I didn't get rid of you. I loved you and I took care of you until you got to the age to have this one. Have your baby no matter what, there's nothing wrong with having a baby. Be proud of it like I was proud of you.' My mama didn't tear me down; she was about the best mother a person ever had."

Unlike many other societies, black women in The Flats feel few if any restrictions about childbearing. Unmarried black women, young and old, are eligible to bear children, and frequently women bearing their first children are quite young.

Excerpted from *All Our Kin, Strategies for Survival in a Black Community* by Carol B. Stack. © 1974, Harper & Row Publishers.

A girl who gives birth as a teenager frequently does not raise and nurture her first-born child. While she may share the same room and household with her baby, her mother, her mother's sister, and her older sister will care for the child and become the child's "mama." This same young woman may actively become a "mama" to a second child she gives birth to a year or two later. When, for example, a grandmother, aunt, or great-aunt "takes a child" from his natural mother, acquired parenthood often lasts throughout the child's lifetime. Although a child kept by a close female relative knows who his mother is, his "mama" is the woman who "raised him up." Young mothers and their firstborn daughters are often raised as sisters, and lasting ties are established between these mothers and their daughters. A child being raised by his grandmother may later become playmates with his half siblings who are his age, but he does not share the same claims and duties and affective ties toward his natural mother.

A young mother is not necessarily considered emotionally ready to nurture a child: for example, a grandmother and other close relatives of Clover Greer, Viola Jackson's neighbor, decided that Clover was not carrying out her parental duties. Nineteen when her first child, Christine, was born, Clover explains, "I really was wild in those days, out on the town all hours of the night, and every night and weekend I layed my girl on my mother. I wasn't living home at the time, but Mama kept Christine most of the time. One day Mama up and said I was making a fool of her, and she was going to take my child and raise her right. She said I was immature and that I had no business being a mother the way I was acting. All my mama's people agreed, and there was nothing I could do. So Mama took my child. Christine is six years old now. About a year ago I got married to Gus and we wanted to take Christine back. My baby, Earl, was living with us anyway. Mama blew up and told everyone how I was doing her. She dragged my name in the mud and people talked so much it really hurt." Gossip and pressure from close kin and friends made it possible for the grandmother to exercise her grandparental right to take the child into her home and raise her there.

In the eyes of the community, a young mother who does not perform her duties has not validated her claim to parenthood. The person who actively becomes the "mama" acquires the major cluster of parental rights accorded to the mothers in The Flats. In effect, a young mother transfers some of her claims to motherhood without surrendering all of her rights to the child.

Fatherhood

People in The Flats expect to change friends frequently through a series of encounters. Demands on friendships are great, but social-economic pressures on male-female relationships are even greater. Therefore, relationships between young, unmarried, childbearing adults are highly unstable. Some men and child-bearing women in The Flats establish long-term liaisons with one another, some maintain sexual unions with more than one person at a time, and still others get married. However, very few women in The Flats are married before they have given birth to one or more children. When a man and woman have a sexual partnership, especially if the woman has no other on-going sexual relationships, the man is identified with children born to the woman. Short-term sexual partnerships are recognized by the community even if a man and woman do not share a household and domestic responsibilities. The offspring of these unions are publicly accepted by the community; a child's existence seems to legitimize the child in the eyes of the community.

But the fact of birth does not provide a child with a chain of socially recognized relatives through his father. Even though the community accepts the child, the culturally significant issue in terms of the economics of everyday life is whether any man involved in a sexual relationship with a woman provides a newborn child with kinship affiliations. A child is eligible to participate in the personal kinship network of his father if the father becomes an immediate sponsor of a child's kinship network.

When an unmarried woman in The Flats becomes pregnant or gives birth to a child, she often tells her friends and kin who the father is. The man has a number of alternatives open to him. Sometimes he publicly denies paternity by implying to his friends and kin that the father could be any number of other men, and that he had "information that she is no good and has been creeping on him all along." The community generally accepts the man's denial of paternity. It is doubtful that under these conditions this man and his kin would assume any parental duties anyway. The man's failure to assent to being the father leaves the child without recognized kinship ties through a male. Subsequent "boyfriends" of the mother may assume the paternal duties of discipline and support and receive the child's affection, but all paternal rights in the child belong to the mother and her kinsmen. The pattern whereby black children derive all their kin through females has been stereotyped and exaggerated in the literature on black families. In fact, fathers in The Flats openly recognized 484 (69 percent) of 700 children included in my AFDC survey.

The second alternative open to a man involved in a sexual relationship with a mother is to acknowledge openly that he is responsible. The father can acknowledge the child by saying "he own it," by telling his people and friends that he is the father, by paying part of the hospital bill, or by bringing milk and diapers to the mother after the birth of the child. The parents may not have ever shared a household and the affective and sexual relationship between them may have ended before the birth of the child.

The more a father and his kin help a mother and her child, the more completely they validate their parental rights. However, since many black American males have little or no access to steady and productive employment, they are rarely able to support and maintain their families. This has made it practically impossible for most poor black males to assume financial duties as parents. People in The Flats believe a father should help his child, but they know that a mother cannot count on his help. But, the community expects a father's kin to help out. The black male who does not actively become a "daddy," but acknowledges a child and offers his kin to that child, in effect, is validating his rights. Often it is the father's kin who activate the claim to rights in the child.

Fatherhood, then, belongs to the presumed genitor if he, or others for him, choose to validate his claim. Kinship through males is reckoned through a chain of social recognition. If the father fails to do anything beyond merely acknowledging the child, he surrenders most of his rights, and this claim can be shared or transferred to the father's kin, whose claim becomes strengthened if they actively participate as essential kin. By failing to perform parental duties the father retains practically no rights in his child, although his kin retain rights if they assume active responsibility.

By validating his claim as a parent the father offers the child his blood relatives and their husbands and wives as the child's kin—an inheritance so to speak. As long as the father acknowledges his parental entitlement, his relatives, especially his mother and sisters, consider themselves kin to the child and therefore responsible for him. Even when the mother "takes up with another man," her child retains the original set of kin gained through the father who sponsored him.

A nonparticipating father also shares some of his rights and duties with his child's mother's current boyfriend or husband. When a man and woman have a continuing sexual relationship, even if the man is not the father of any of the woman's children, he is expected by the mother and the community to share some of the parental duties of discipline, support, and affection.

Child-Keeping

The black community has long recognized the problems and difficulties that all mothers in poverty share. Shared parental responsibility among kin has been the response. The families I knew in The Flats told me of many circumstances that required co-resident kinsmen to take care of one another's children or situations that required children to stay in a household that did not include their biological parents.

People in The Flats often regard child-keeping as part of the flux and elasticity of residence. The expansion and contraction of households, and the successive recombinations of kinsmen residing together, require adults to care for the children residing in their household. As households shift, rights and responsibilities with regard to children are shared. Those women and men who temporarily assume the kinship obligation to care for a child, fostering the child indefinitely, acquire the major cluster of rights and duties ideally associated with "parenthood."

Within a network of cooperating kinsmen, there may be three or more adults with whom, in turn, a child resides. In this cycle of residence changes, the size of the dwelling, employment, and many other factors determine where children sleep. Although patterns of eating, visiting, and child care may bring mothers and their children together for most of the day, the adults immediately responsible for the child change with the child's residence. The residence patterns of children in The Flats have structural implications for both the ways in which rights in children distribute socially and also the criteria by which persons are entitled to parental roles.

From the point of view of the children, there may be a number of women who act as "mothers" toward them; some just slightly older than the children themselves. A woman who intermittently raises a sister's or a niece's or a cousin's child regards their offspring as much her grandchildren as children born to her own son and daughter.

The number of people who can assume appropriate behaviors ideally associated with parental and grandparental roles is increased to include close kinsmen and friends. Consequently, the kin terms "mother," "father," "grandmother," and the like are not necessarily appropriate labels for describing the social roles. Children may retain ties with their parents and siblings and at the same time establish comparable relationships with other kinsmen. There is even a larger number of friends and relatives who may request a hug and kiss, "a little sugar," from children they watch grow up. But they do not consistently assume parental roles toward those children. Parental role behavior is a composite of many behavior patterns (Keesing 1969) and these rights and duties can be shared or transferred to other individuals.

Bonds of obligation, alliance, and dependence among kinsmen are created and strengthened in a variety of ways. Goods and services are the main currency exchanged among cooperating kinsmen. Children too may be transferred back and forth, "borrowed" or "loaned." It is not uncommon for individuals to talk about their residence away from their mother as a fact over which she had little or no control. For example, kin may insist upon "taking" a child to help out. Betty Simpson's story repeats itself with her own daughter. "My mother already had three children when I was born. She had been raised by her maternal great-aunt. After I was born my mother's great-aunt insisted on taking me to help my mother out. I stayed there after my mother got married and moved to The Flats. I wanted to move there too, but my 'mama' didn't want to give me up and my mother didn't want to fight with her. When I was fourteen I left anyway and my mother took me in when my youngest daughter got polio my mother insisted on taking her. I got a job and lived nearby with my son. My mother raised my little girl until my girl died."

A mother may request or require kin to keep one of her children. An offer to keep the child of a kinsman has a variety of implications for child givers and receivers. It may be that the mother has come upon hard times and desperately wants her close kinsmen to temporarily assume responsibility for her children. Kinsmen rarely refuse such requests to keep one another's children. Likewise they recognize the right of kin to request children to raise away from their own parents. Individuals allow kinsmen to create alliances and obligations toward one another, obligations which may be called upon in the future.

It might appear that the events described above contribute to a rather random relocation of individuals in dwellings, and a random distribution of the rights individuals acquire in children. But this is not the case. Individuals constantly face the reality that they may need the help of kin for themselves and their children. As a result they anticipate these needs, and from year to year they have a very clear notion of which kinsmen would be willing to help. Their appraisal is simple because it is an outcome of calculated exchanges of goods and services between kinsmen. Consequently, residence patterns and the dispersing of children in households of kin are not haphazard.

Reflections

1. In this system, who has the right to be called "mama"?

2. How does a father validate his paternal rights?

3. What are the strengths and weaknesses of this informal system? What role does poverty play in it?

4. Reflecting on your own childhood, how might you feel in such a community?

37

This article summarizes psychiatrist Dr. Robert Beavers' five basic components of family success. Using these, he believes, families can enjoy each other more and consequently become healthy. Think about your own family as you read the article.

Of Course You Love Them—But Do You Enjoy Them?

By Anita Shreve

Today, with both fathers and mothers in most families working outside the home, children grow up exposed to a large number of authority figures beyond the family—day-care instructors, teachers, baby sitters. Furthermore, contemporary society places great demands on its people to be decision makers and autonomous—to grow up. For these reasons, according to Dr. Robert Beavers, clinical professor of psychiatry at the University of Texas Health Center, in Dallas, and director of the Southwest Family Institute, parents would do well to invite children to be full participants and partners in family life. As for the question "How well does my family function?" Dr. Beavers suggests that the only question you have to ask yourself is: "Am I enjoying my family?" This, he says, is the acid test for a healthy family.

The ability to take pleasure in one's family, and in oneself as a member of that family, is, according to the family specialist, the most important goal worth striving for. If the answer to "Do I like living here?" is yes, parents should relax and not put their families under the microscope. "But if the answer is no," Dr. Beavers says, "ask yourself another question: 'What is the thing that would make me enjoy this family more?' You have to ask yourself that because you can't acquire something you can't first imagine."

To help parents get more enjoyment from their families—and in turn make their families healthier—Dr. Beavers outlines five fundamental components of—or steps toward—family success. When you pay attention to these five components of a healthy family, he believes, family enjoyment will develop as a natural consequence.

1. It Starts with the Marriage

A strong intimate, sexual, happy marriage in which the parents are truly good friends sets the tone for a happy family life. Research indicates that parents of healthy families share power, have a good sexual relationship and are separate individuals despite strong emotional ties. Most important, they act as a team.

"You have to have an intimate friend," says Dr. Beavers, "someone you can trust and with whom you can talk about the kids. This is absolutely vital and it's just as important for single parents too. They need to find someone—perhaps an older woman who has already had her children—to talk with about raising a family."

As with family success, assessing whether or not you have a good marriage starts with a simple question: "Do I like living with this person?"

"If you're not enjoying your marriage," Dr. Beavers says, "try to identify what it is that you need. Try to complete the sentence 'I want _____.' Once you actually imagine what it is you need, you can set about trying to get it."

Although a happy marriage is optimal, it is possible to achieve a good family life without one—but only if parents act as a team. "I've seen families in which the parents had a terrible marriage but ran a 'good business,'" says Dr. Beavers. "They could agree on how to run the household even though they didn't have intimacy themselves."

Finally, the psychiatrist warns, it's important for a parent not to team up with a child against the other parent. Such divisiveness leads inevitably to

© 1983 by Anita Shreve.

jealousy, anger and a destruction of whole-family intimacy.

2. Learn to Enjoy Your Children

The scene is a familiar one. During a dinner party at the home of friends of his parents, two-and-a-half-year-old Bill alternately stands on the table, spills his milk, refuses to eat anything that is put in front of him and catapults his peas from his own plate into his hostess's lap with uncanny accuracy. The entire dinner is punctuated with apologies and exclamations of helplessness and dismay. With the possible exception of Billy, everyone is miserable. What has gone wrong?

According to Dr. Beavers, Billy's parents have made themselves martyrs by refusing to lay down basic rules of behavior that would make their own lives easier and would allow them to enjoy their son. "There have to be certain basic rules designed expressly for Mom and Dad's comfort and based on the parents' desire to enjoy themselves. After all, if you don't have rights of your own, you can't very well give them to someone else."

"Thou shalt enjoy thy children" is the first commandment of parenting. Assuming a long-suffering air of martyrdom while the children run wild won't allow parents to take pleasure in them. "In cases in which they haven't set limits, the parents alternate between being 'dutiful' and 'going bananas,'" says Dr. Beavers. "Family rules need to be clear, agreed on, and enforced to give parents and children a secure and pleasant home."

3. Lead but Don't Control

Setting basic rules for young children, however, does not mean the same thing as trying to control their thoughts and feelings or ruling like a dictator. Research indicates that although power in healthy families rests securely with the parents, they do not rule as authoritarians. Instead they provide easy leadership, listening to children's opinions and feelings with respect and trying to solve disputes by clarifying issues and negotiating compromises.

Dr. Beavers, who describes himself as a "parent emeritus" and says he has been "consulted" rather than obeyed since his children were small, explains: "Parents cannot—and therefore shouldn't try to—control the thoughts and feelings of their children. Accepting that eliminates two-thirds of the problem of parenting."

He gives the example of the child who has bad grades in school: A parent cannot force a child to want to make good grades, but he or she can say, "We're going to have two hours in which there will be no distractions." This allows the child to decide for himself to do better in school by using that time to study.

Dr. Beavers believes working together as parents in developing and maintaining family rules is quite different from attempting to control a child's goals. "Efforts as controlling a child's directions, successes, loves and hates are always frustrating to both parent and child," he points out.

The benefits of relaxed leadership are enormous, for it allows the members of the family to be intimate. Or, put another way, you can't have intimacy if one person has power and the other doesn't. This can be seen most easily in a marriage. If all the power in the marriage rests with the husband and the wife is never consulted, respected, or allowed to make decisions, there can be no real intimacy between them. The same is true with children. If the parent demands unwavering obedience from the child without ever considering the child's perceptions, the relationship will lack intimacy.

4. Learn How to Negotiate

Over the years, Dr. Beavers has learned to identify a healthy family on the basis of two criteria: (1) Do the members of a family feel good? ("Does it feel good to be around the others and do they have an air of feeling good themselves?"); and (2) Do they know how to negotiate?

"Healthy families know how this is done," he says. "In fact, negotiation may be the single most important factor in developing healthy relationships. Conflict is ever-present. But if a family can dicker its way through these conflicts, it's probably in pretty good shape."

Negotiating starts with an ability to resolve one's own feelings. Each of us has mixed attitudes about many issues. But if we're able to resolve our ambivalence and then communicate clearly these resolved feelings—at the same time allowing other family members to do this too—then there is a fairly sound basis for good negotiating.

An illustration of this is a typical family trying to decide what to do on a free Sunday afternoon. Dad wants to mow the lawn. Michael, who is eight years old, wants the family to go bike riding around the lake. Mom wants to go to the lake too, and thinks that two-year-old Jennifer will enjoy riding in the baby seat on the back of her bike. But she's ambivalent because she knows if the lawn doesn't get done, it will be a whole week before Dad can return to it. Mom resolves her ambivalence this way: "I'm torn, but I think a family outing is what we all need right now. Therefore I'll lobby for that." The family begins to negotiate. Dad comes on strong for mowing the lawn. Michael is equally determined to go bike riding. Mom, her feelings resolved, states her position but sees a way to bargain. "Let's take a picnic to the lake, eat lunch, ride for half an hour and then come home. By that time, the grass will be dry and Dad can mow the lawn." After a little more discussion, everyone agrees.

Members of healthy families express their feelings—whether they be anger, sadness, love or fear—because the family ambiance is basically one of understanding and warmth. But the idea of totally honest and open communication in a family is a myth, admits Dr. Beavers. "Only when your child is three years old will he or she be completely open and clear—able to say 'I hate you Mommy' as easily as 'I love you.' Part of being in an intimate relationship is learning to keep certain things to oneself."

5. Learn to Let Go

The usual image of the happy, healthy family is one in which all the family members enjoy doing the same things together—like camping, gardening, or just reading the paper on a rainy Sunday afternoon. But members of healthy families have also learned to be separate individuals, each with an ability to think for himself and to make decisions for himself. The ability to *choose* how to be or how to feel is called *autonomy*, and it is the most treasured asset a parent can give to a child.

Let's take the example of a nine-year-old girl named Joanne who has just formed a friendship with a new girl in school. The new friend is of a racial background different from Joanne's, and although her parents know better, they can't help but express reservations and concern based on lingering prejudices with which they themselves grew up. Joanne's affection for the girl is quite real, however, and the two have discovered they like many of the same activities. Because Joanne's parents have fostered in her an ability to think for herself, she feels good about her newly found friendship. If Joanne's parents had denied her the ability to think for herself and accept responsibility for her actions, she might have acquiesced to their nameless fears and prejudices and given up the friend—or kept her but felt uncomfortable and guilty about doing so.

"Autonomy must be developed," says Dr. Beavers. "One must learn what one feels and thinks and must be able to make decisions and accept responsibility for these decisions. For a parent, the job is to render oneself unnecessary, to see the family as a dissolving unit, leaving the couple with the love ties with which they began but allowing the children to move on. Thus parents have given their children autonomy."

Although it may at first seem strange and uncomfortable to think of the ultimate goal of the family as that of dissolving, this process is necessary in order for the children to grow up to be adults. Parents first need to have a firm sense of their own identity to help the children develop theirs. (Remember that you can't give to someone else what you don't have yourself.) A parent is a person first—and if this is made clear, it allows the child to realize what a grownup can be—a separate individual with unique thoughts and feelings and perspectives. This process can then be facilitated by some of the parenting skills mentioned earlier—setting basic rules, learning how to negotiate, inviting your children into the family as participants and encouraging the expression of feelings.

Being separate does not, however, mean being distant. According to Dr. Beavers, there is more autonomy in close-knit families than in distant ones. "In close-knit families the children feel the roots but know that those roots allow for some breathing space—allow them to say 'I choose.' To be close, you must first learn to be separate. You can't tolerate intimacy if you're not separate, because you fear that you'll be swallowed up."

Although following these five steps may help you have a happier family life, it will be obvious to any parent that the steps are interrelated and complex and that family interaction cannot be easily divided into categories. Negotiating, for instance, is dependent on the individual family members' ability to develop autonomy; skill in negotiating with children may be directly related to how well the parents can negotiate and how happy their marriage is; establishing casual leadership probably will work only if everyone in the family already respects the basic family rules.

And certainly no one—especially Dr. Beavers—wants to see a family straining to conform to a rigid system of family life, especially if doing so doesn't make the members happy. In fact, the psychiatrist is the first to suggest that each couple try to find their own way of making a family. "Every marriage is like every new pregnancy," he says. "In the same way that there are infinite possibilities in the unborn, there are also infinite possibilities for a couple to create a family. When two people come together they hold between them the potential to make something unique—a way of being that has never been before. I like to think that instead of conforming to the way of making a healthy family, they will say, 'We can do it our way.'"

Reflections

1. When reflecting about your own family, do you see areas where the potential for enjoyment could be further developed? How might this happen?

2. Which issue is your family's greatest strength? Weakness?

3. Did you enjoy being part of your family?

In this reading, the author describes how he and his family began to incorporate a new tradition, Kwanzaa, into their lives.

Kwanzaa: A "New" Tradition for African Americans

By Eric V. Copage

I was never a holiday kind of guy.

Perhaps it was because we observed few holiday rituals of any kind. Although we put up a Christmas tree every year, there was no ceremony to it—no drinking of eggnog or listening to carols while hanging ornaments. To me, the tree seemed more or less like another piece of furniture.

Over the past few years, however, the holiday season has taken on a new meaning for me as my family sits at the dinner table the week after Christmas, to celebrate Kwanzaa. This cultural observance for black Americans and other of African descent was created in 1966 by Maulana "Ron" Karenga, now chairman of black studies at California State University, Long Beach.

Kwanzaa means "first fruits of the harvest" in Swahili, but there is no festival of that name in any African society. Karenga synthesized elements from many African harvest festivals to create a unique celebration that is now observed in some way by more than 5 million Americans.

When I first told my wife I was thinking about observing Kwanzaa, she barred the way to our attic and said she'd never chuck our Christmas tree lights and antique ornaments. I told her that wouldn't be necessary. Kwanzaa, which runs from Dec. 26 to New Year's Day, does not replace Christmas and is not a religious holiday. It is a time to focus on Africa and African-inspired culture and to reinforce a value system that goes back for generations.

I was introduced to Kwanzaa by chance in late December a few years back. I was visiting the American Museum of Natural History when I heard the sibilant sound of African rattles. It was coming from a dance performance, part of the Kwanzaa celebration that has been held annually at the museum since 1978.

The holiday didn't make much of impression on me then, but I returned to it after the birth of my son. I wanted him to have a three-dimensional sense of his African heritage. I wanted him to experience the pride of learning about the sublime Russian poet Aleksandr Pushkin, the extraordinary composer Duke Ellington and Alexander Dumas, author of "The Three Musketeers." I wanted him to learn about the West African medieval empires and about African explorers and inventors.

I wanted him to understand that through tenacity, hard work and purposefulness, blacks have flourished as well as survived. I wanted to train him to look for opportunity and prepare for it. And I wanted to have a forum for showing him examples of past successes, and for showing him that those people inevitably gave back to the black community in particular and to the general community in which they lived.

I thought about my goals for my son and decided that Kwanzaa was the best lens through which to view the landscape of the African diaspora and the lessons it has to teach. Because it is only one week long and because it climaxes with a glorious feast, Kwanzaa has an intensity and focus that provide the perfect atmosphere for my son to experience the joys of being black. Kwanzaa also has the celebratory aspect that will provide memories for him—and now my daughter—to savor as adults and to pass on to their children.

When my family lights the black, red and green Kwanzaa candles the last week of December, we do so with millions of other black Americans around the nation. Major community celebrations

From Kwanzaa: A Celebration of Culture and Cooking. Copyright © 1991 by Eric V. Copage. By permission of William Morrow and Company, Inc.

are held in just about every city that has a substantial black population.

Our first meal consists of dishes that bring forth sweet memories of my childhood. I remember helping my grandmother make collard greens and sorting the black-eyed peas for the Hoppin' John she cooked. I remember visiting West Africa when I was 18 and tasting the spicy tingle of peanut soup for the first time. And, of course, there was a lifetime of corn bread.

Like my family's Christmas, our Kwanzaa tends to be a small celebration among our nuclear family of four. After our Kwanzaa meal, I relate the biography of a black man or woman or tell of a black folktale, myth or historical event that illustrates one of the seven principles.

During my family's Kwanzaa we don't drink from the unity cup, but rather pour a small libation into it and leave it in the center of the table. And we prefer to use freestanding candles instead of putting them in a kinara.

That may change, however, with this coming Kwanzaa or the Kwanzaa after that. Since the celebration of Kwanzaa is destined to transform further as African-Americans calibrate their observances to what they are comfortable with and to the needs of their individual families.

It took several years for me to feel comfortable saying "Happy Kwanzaa." But I think that is only natural. Any holiday, and the rituals that go with it, derives its symbolic and social power from its cultural context. It takes time for a cultural context to crystallize.

Starting Your Own Tradition

If you want to adhere strictly to the Kwanzaa program as conceived by Maulana "Ron" Karenga, here is what you need and what they mean:

- **Mazao:** Fruits and vegetables, which stand for the product of unified effort.
- **Mkeka:** A straw place mat, which represents the reverence for tradition.
- **Vivunzi:** An ear of corn for each child in the family.
- **Zawadi:** Simple gifts, preferably related to education or to things African or African-influenced.
- **Kikombe Cha Umoja:** A communal cup for libation (I like to look at this as a kind of homage to past, present and future black Americans).
- **Kinara:** A seven-branched candle holder, which symbolizes the continent and peoples of Africa.
- **Mishumaa Saba:** Seven candles, each of which symbolizes one of the Nguzo Saba, or seven principles, that black Americans should live by on a daily basis and that are reinforced during Kwanzaa.

On each day of the Kwanzaa celebration, a family member lights a candle, then discuss one of those seven principles. The principles, along with Karenga's elucidation of them in 1963, are:

1. **Umoja (Unity):** To strive for and maintain unity in the family, community, nation and race.
2. **Kujichagulia (Self-Determination):** To define ourselves, name ourselves, create for ourselves and speak for ourselves instead of being defined, named, created for and spoken for by others.
3. **Ujima (Collective Work and Responsibility):** To build and maintain our community together, and to make our sisters' and brothers' problems our problems, and to solve them together.
4. **Ujamma (Cooperative Economics):** To build and maintain our own stores, shops and other businesses and to profit from them together.
5. **Nia (Purpose):** To make our collective vocation the building and developing of our community in order to restore our people to their traditional greatness.
6. **Kuumba (Creativity):** To do always as much as we can, in whatever way we can, in order to leave our community more beautiful and beneficial than we inherited it.
7. **Imani (Faith):** To believe with all our heart in our people, our parents, our teachers, our leaders and in the righteousness and victory of our struggle.

The next-to-last day of the holiday, Dec. 31 (many people do this on Jan. 1), is marked by a lavish feast, the Kwanzaa Karamu, which, in keeping with the theme of black unity, may draw on the cuisines of the Caribbean, Africa, and South America.

How to Conduct a Kwanzaa Karamu

In addition to food, the Kwanzaa Karamu is an opportunity for a confetti storm of cultural expression—dance and music, readings, remembrances. Here is how to conduct a Karamu. Each of the following is punctuated by some form of artistic expression of African or African-American heritage.

- **Kukaribisha (Welcoming):** Introductory remarks and recognition of distinguished guests and elders.
- **Kukumbuka (Remembering):** Reflections of a man, a woman and a child.
- **Kuchungúza Tena Na Kutoa Ahadi Tena (Reassessment and Recommitment):** Introduction of distinguished guest lecturer and short talk.
- **Kushangilia (Rejoicing):** Includes the pouring of a libation from the communal cup, a libation statement, the calling of names and family ancestors and black heroes, drums and the feast.
- **A farewell statement.**

Reflections

1. What holidays are important in your family? Will you pass them on to your children?

2. Would you like to change the way they are celebrated or to create new rituals?